51. K.V. Novozhilov (Chief editor): *Microbiological Methods for Biological Control of Pests of Agricultural Crops*
52. K.I. Rossinskii (editor): *Dynamics and Thermal Regimes of Rivers*
53. K.V. Gnedin: *Operating Conditions and Hydraulics of Horizontal Settling Tanks*
54. G.A. Zakladnoi & V.F. Ratanova: *Stored-grain Pests and Their Control*
55. Ts.E. Mirtskhulava: *Reliability of Hydro-reclamation Installations*
56. Ia. S. Ageikin: *Off-the-road Mobility of Automobiles*
57. A.A. Kmito & Yu.A. Sklyarov: *Pyrheliometry*
58. N.S. Motsonelidze: *Stability and Seismic Resistance of Buttress Dams*
59. Ia.S. Ageikin: *Off-the-road Wheeled and Combined Traction Devices*
60. Iu.N. Fadeev & K.V. Novozhilov: *Integrated Plant Protection*
61. N.A. Izyumova: *Parasitic Fauna of Reservoir Fishes of the USSR and Its Evolution*
62. O.A. Skarlato (Editor-in-Chief): *Investigation of Monogeneans in the USSR*
63. A.I. Ivanov: *Alfalfa*
64. Z.S. Bronshtein: *Fresh-water Ostracoda*
65. M.G. Chukhrii: *An Atlas of the Ultrastructure of Viruses of Lepidopteran Pests of Plants*
66. E.S. Bosoi et al.: *Theory, Construction and Calculations of Agricultural Machines*, Volume 1
67. G.A. Avsyuk (Editor-in-Chief): *Data of Glaciological Studies*
68. G.A. Mchedlidze: *Fossil Cetacea of the Caucasus*
69. A.M. Akramkhodzhaev: *Geology and Exploration of Oil- and Gas-bearing Ancient Deltas*
70. N.M. Berezina & D.A. Kaushanskii: *Presowing Irradiation of Plant Seeds*
71. G.U. Lindberg & Z.V. Krasyukova: *Fishes of the Sea of Japan and the Adjacent Deltas*
72. N.I. Plotnikov & I.I. Roginets: *Hydrogeology of Ore Deposits*
73. A.V. Balushkin: *Morphological Bases of the Systematics and Phylogeny of the Nototheniid Fishes*
74. E.Z. Pozin et al.: *Coal Cutting by Winning Machines*
75. S.S. Shul'man: *Myxosporidia of the USSR*
76. G.N. Gogonenkov: *Seismic Prospecting for Sedimentary Formations*
77. I.M. Batugina & I.M. Petukhov: *Geodynamic Zoning of Mineral Deposits for Planning and Exploitation of Mines*
78. I.I. Abramovich & I.G. Klushin: *Geodynamics and Metallogeny of Folded Belts*
79. M.V. Mina: *Microevolution of Fishes*
80. K.V. Konyaev: *Spectral Analysis of Physical Oceanographic Data*
81. A.I. Tseitlin & A.A. Kusainov: *Role of Internal Friction in Dynamic Analysis of Structures*
82. E.A. Kozlov: *Migration in Seismic Prospecting*
83. E.S. Bosoi et al.: *Theory, Construction and Calculations of Agricultural Machines*, Volume 2
84. B.B. Kudryashov and A.M. Yakovlev: *Drilling in the Permafrost*
85. T.T. Klubova: *Clayey Reservoirs of Oil and Gas*
86. G.I. Amurskii et al.: *Remote-sensing Methods in Studying Tectonic Fractures in Oil- and Gas-bearing Formations*
87. A.V. Razvalyaev: *Continental Rift Formation and Its Prehistory*
88. V.A. Ivovich & L.N. Pokrovskii: *Dynamic Analysis of Suspended Roof Systems*
89. N.P. Kozlov (Technical Editor): *Earth's Nature from Space*
90. M.M. Grachevskii & A.S. Kravchuk: *Hydrocarbon Potential of Oceanic Reefs of the World*
91. K.V. Mikhailov et al.: *Polymer Concretes and Their Structural Uses*
92. D.S. Orlov: *Soil Chemistry*
93. L.S. Belousova & L.V. Denisova: *Rare Plants of the World*
94. T.I. Frolova et al.: *Magmatism and Transformation of Active Areas of the Earth's Crust*
95. Z.G. Ter-Martirosyan: *Rheological Parameters of Soils and Design of Foundations*

RHEOLOGICAL PARAMETERS OF SOILS AND DESIGN OF FOUNDATIONS

RHEOLOGICAL PARAMETERS OF SOILS AND DESIGN OF FOUNDATIONS

Z.G. TER-MARTIROSYAN

RUSSIAN TRANSLATIONS SERIES

95

A.A. BALKEMA PUBLISHERS/ROTTERDAM/BROOKFIELD
1992

Translation of: *Rheologicheskii parametri gruntov i rascheti osnovanii sooruzhenii* Nedra, Moscow, 1990

Translator   : Dr. N.K. Mehta
Technical Editor   : Dr. G. Venkatachalam
General Editor   : Ms. Margaret Majithia

ISBN 90 5410 211 X

Distributed in USA & Canada by: A.A. Balkema Publishers, Old Post Road, Brookfield, VT 05036, USA.

Preface

In design, construction and operation of various structures built under difficult technical and geological conditions, it is necessary to predict the stress-strain state of soil mass, that is interacting with the structure. The reliability and accuracy of this estimate depend upon the real properties of soil taken into consideration and the parameters included in the analysis of the stress-strain state. Rheological parameters of soil in the limit and prelimit states (in terms of strength) are the first concern. In considering soil properties and determining their parameters, one requires complex instrumentation and well-worked-out procedures for conducting laboratory and field tests. On the other hand, an analysis of the stress-strain state may be carried out by numerical methods on computers. A review of recent publications and proceedings of international conferences and seminars revealed that computerisation has been widely introduced in analytical research in the field of applied geomechanics, including soil rheology.

The present book is a generalisation of the work carried out by the author and his students in the field of soil rheology at the V.V. Kuibyshev, Moscow Civil Engineering Institute.

The results of these investigations were used in compiling recommendations for the determination of rheological properties of soils, design norms for bases of the reactor divisions of atomic power plants and analysis of unique, heavy structures, slopes and embankments. These recommendations resulted in a saving of about 3 million roubles.

The concepts and principles of soil rheology were formulated by S.S. Vyalov, M.N. Gol'dshtein, N.N. Maslov, G.I. Ter-Stepanyan, V.A. Florin and N.A. Tsytovich and were later developed by A.Ya. Budin, A.L. Gol'din, L.V. Gorelik, S.E. Grechishchev, Yu.K. Zaretskii, S.P. Meschyan etc. The pressing need to develop soil rheology as a branch of applied geomechanics was emphasised on more than one occasion by N.A. Tsytovich.

The contents of the present book include results of investigations of the rheological properties of soils, methods of determination of rheological parameters of soils and methods of analysis of the stress-strain state of soil mass, taking into account the rheological properties. In a bid to emphasise the distinguishing features of the rheology of clayey soils, the author has been forced to refrain from a detailed description of many important problems of soil mechanics. Due to the same considerations, the discussion of contentious issues in soil rheology vi

has been limited in scope and only the phenomenological fundamentals of the relevant theoretical principles have been enunciated.

The author is grateful to the Honoured Scientist of the Russian Federation, Prof. P.L. Ivanov, D.Sc. for painstakingly reading the manuscript and giving valuable comments for improving the contents of this book. Thanks are also due to I.M. Yudina, Ph.D. for her assistance in preparation of the manuscript and printing of the book.

Contents

1

Rheological Properties of Clayey Soils

1. General Principles and Concepts

The rheological properties of clayey soils always play an important role in the interaction between the soil and the structures and geological environment. This is true in equal measure for creep, relaxation and variation of soil strength in time, i.e., the processes that have a significant effect on the nature of the stress-strain state in a clayey soil mass and the changes that occur in it in time and space, thereby determining the service conditions and stability of the structures.

There are numerous instances of rheological processes in the foundations having resulted in large and non-uniform settlement, considerable displacement of retaining walls from their original position and instability of slopes and embankments. Even from these few examples it becomes evident that a solution to the problems of applied geomechanics is impossible without taking into account the rheological properties of clayey soils.

As clayey soil is a multiphase medium, consisting of soil skeleton (mineral lattice) and pore water with dissolved gases, the rheological processes in clayey soils proceed in different ways, depending on the degree of saturation of the pores with water.

In a clayey soil that is not fully saturated with water and has a degree of saturation $S_r < 0.8$, the rheological process is determined by the creep of soil skeleton because, during deformation of soil, the pore water has no effect on volumetric changes. Therefore, the traditional methods of the rheology of single-phase media can be applied to the study and description of rheological processes in clayey soils.

In water-saturated soils $(S_r > 0.8)$ the rheological process is governed, on the one hand, by the creep of soil skeleton and, on the other, by the compressibility and viscous resistance of pore water to compaction. Obviously, such a process cannot be studied and described by the traditional methods of the rheology of single-phase media. Therefore, a special approach needs to be developed for this purpose. This is particularly important for soils in which deformation is accompanied by formation of excess pore pressure in pore water, followed by its gradual dissipation. In such cases it is necessary to consider the interaction between the phases and the variation of their relative proportions in a unit volume of the soil.

A number of principles have been adopted in soil mechanics to describe this complex process, viz., Terzaghi's principle of effective stresses, Leibinson's principle of the compressibility of pore water, Darcy's principle of laminar seepage, principle of effective modulus etc. However, researchers dealing with this problem often forget that all the aforesaid principles assume a certain idealisation, which, like all idealisations, impose certain constraints. The advantages and drawbacks of the idealisation are manifested more clearly while dealing with specific applied problems.

For instance, the principle of effective stresses allows the equations of the rheology of soils to be employed for description of temporal processes in the soil as a whole and also at various points in it during the state of unstabilised compression (consolidation). It is absolutely clear that erroneous results would be obtained if the equations of continuum mechanics were used instead of the principle of effective stresses for description of temporal processes occurring during deformation of clayey soils.

Clayey soil being a discrete medium, the principle of homogeneity and continuity is also a necessary and compulsory precondition for application of the methods of continuum mechanics to clayey soils. It is assumed that clayey soil is isotropic and the elementary volume of that soil whose deformation is under study is bigger than the soil grains. It is further assumed that only these units contain anisotropic dispersed elements (mineral particles) of all possible orientations. In view of the above assumptions, the soil volumes considered for rheological studies should be minimal in size, but should contain an adequate quantity of elementary soil constituents and display all the properties characteristic of a soil medium. Consequently, the concept of point in a solid body has a specific implied meaning when it is applied to describe the stress-strain state in a soil mass, i.e., it represents a certain volume around the given point. Therefore, the stress or strain at a point of the mass should be understood to represent the average value in a small volume around the given point. For these reasons, when soil specimens are taken from the mass for testing, their dimensions should be selected with due consideration to the scale of elementary volume. It may be mentioned, however, that the rheological properties of soil determined from results of laboratory tests are not always able to establish the dynamics of the temporal processes in the soil mass, particularly if the soil is water saturated. In the latter case, the rheological process is additionally governed by the rheological properties of the soil skeleton and consolidation due to seepage.

Degree of saturation and permeability play a significant role in the build up of the stress-strain state in clayey soil mass and the changes that occur in it. Even the presence of 1% gas bubbles in clayey soil leads to a significant change in the build up of a stress-strain state in the soil mass in the initial stage as well as during consolidation under seepage.

The degree to which the creep behaviour of the skeleton and the filtration behaviour of the soil affect the rheological process in a soil mass depends on the coefficient of permeability (k), creep of skeleton and length of the seepage path. To evaluate the contribution of one or the other factor to the rheological process as a whole, the author [17] has introduced a non-dimensional parameter that describes the integral viscosity of soil mass, including the viscosity of the skeleton and pore water when it is squeezed out of the pores:

$$\mu_c = \delta(L^2/c_v), \qquad \ldots (1.1)$$

where δ is the parameter describing the creep of soil skeleton $(1/s)$, L is the maximum length of drainage path (cm) and c_v is the coefficient of consolidation (cm^2/s).

For soils in which $k \geq 10^{-5}$ cm/s, at $\mu_c \leq 0.001$ the rate of volumetric strain is determined mainly by creep of skeleton. In such cases it suffices to consider only creep of soil skeleton in studying the temporal processes; the consolidation aspect may be ignored. For soils with a small value of coefficient of permeability $k \leq 10^{-8}$ cm/s, at $\mu_c \geq 10$, due to the large size of the soil mass or large viscosity of the skeleton, the temporal process in the mass is determined mainly by the viscous resistance of pore water to squeezing. Consequently, in such cases it is not obligatory to consider creep of skeleton while studying the change in stress-strain state with time.

Hence, in each particular case, a preliminary estimation of parameter μ_c is essential in deciding whether creep of soil skeleton should be considered while studying the settlement of foundations of structures due to volumetric deformation of the skeleton. Regarding settlement of foundation due to shear deformation of the skeleton, it is obvious that in this case the rheological properties of the skeleton must be taken into account.

The rheology of soils in application to analysis of foundations of structures has followed such a course of development that initially studies were devoted to creep deformation under shear. In these studies the interaction of pore water with the skeleton was not considered because it did not have an effect on the shearing process due to the absence of volumetric strain. This approach was adopted in the study of landslides and horizontal displacements of the foundations of hydrotechnical structures. Later, with the progress achieved in improved methods of analysis of soil foundations and structures based on the second limit state, it became necessary to study and predict the creep deformation occurring in the course of soil consolidation.

At present, the rheology of clayey soils is developing in both the above-mentioned directions. The stress-strain state of soil is studied by taking into account the shearing as well as volumetric deformation of the skeleton. Further, the compressibility of pore water and its expulsion are also considered. Also, considerable emphasis is given to the effect of density (dampness) and the granulometric and mineral composition of soil on its rheological properties.

Soil rheology is based on the results of experimental investigations, i.e., it is a phenomenological science. In other words, it deals with the macrolevel study and description of externally observable temporal processes which can be recorded by means of measuring equipment. On the whole, soil is assumed to have idealised properties such as elasticity, plasticity, creep and their combinations. The solutions obtained on the basis of experimental studies are found to yield fairly satisfactory results in a given range of stresses and accord well with field observations.

A more general approach would be to establish the laws governing the deformation of soil in time by considering the rheological process at the microlevel. The mechanism of development of the rheological process in clayey soil at the microlevel has a complex, physicochemical character. This process is governed, on the one hand, by the interaction between mineral particles and bonded water and, on the other, by the interaction between soil skeleton and free water. It is not yet possible to consider this mechanism while studying the behaviour of clayey soils, as there are no effective methods for describing the rheological properties of soil on the basis of microlevel studies. Nevertheless, a knowledge of the mechanism of interaction of mineral particles in clayey soil is helpful in explaining the rheological process at the macrolevel. But before dealing with this problem in depth, it is necessary to present an overview of the behaviour of clayey soils.

2. Essential Information on Clayey Soils

Clayey soils are the most widely prevailing sedimentary deposits and are found mostly in the upper layers of the earth's crust. Therefore, while tackling various engineering problems one often has to deal first with the interaction between clayey soils and the structures. There is considerable difference between the deformability and strength of clayey soils in comparison with those of sandy soil, rocky deposits and rock, particularly if the time factor is taken into consideration. Therefore, difficulties arise in solving practical problems when clay is used as the foundation material or construction material for various structures.

The mechanical properties of clayey soils, including the rheological properties, largely depend on the mineralogical and granulometric composition, the proportion of solid (mineral), liquid and gaseous components and also the nature of interaction between these components. For instance, in highly dispersed clays, the coagulational bond dominates over a wide range of density and direct contact between the particles is practically absent. These structures display all the properties of solids and liquids. Clayey soil may be represented as a composite material, consisting of a mineral skeleton, liquid and gas which fill the pores. Obviously, the properties of such a composite are governed by the ratio between the skeleton, liquid and gas and the manner in which these components interact. Here difficulties arise with regard to the term "soil skeleton" because the skeleton is ascribed the properties of a solid, i.e., the soil as a whole. The water stored in the voids or the so-called rigidly bound water is also attached to the skeleton. However, this definition of soil skeleton has not been confirmed experimentally.

The rheological properties of clayey soils vary over a wide range, from those of weak water-saturated silt at one extreme to those of unsaturated clays of hard consistency at the other. While classifying clayey soils special attention should be paid to the determination of the degree of saturation, as it is an important indicator of both quantitative and qualitative parameters of soil.

Depending on the coefficient of permeability, clayey soils may be classified as previous ($k > 10^{-5}$ cm/s), poorly pervious ($10^{-5} > k > 10^{-9}$ cm/s) and almost impervious ($k < 10^{-9}$ cm/s).

Clayey soils also include collapsible loess soils and swelling soils that are structurally unstable. The mechanical processes in these soils are distinguished by the fact that their natural structure is unstable under the simultaneous action of force and moisture fields. Additional moisture has a plasticising (softening) effect on the soil skeleton due to weakening of the cementing bonds and reduction of the overall resistance to shearing. In view of the foregoing, permeability is one of the main properties of these soils.

It follows from the foregoing discussion that clayey soils differ significantly from each other in terms of mechanical as well as physical properties. Consequently, while solving practical problems, it is essential to differentiate between various clayey soils and to describe the rheological processes occurring in a clayey soil mass by taking into account the specific features of the given soil.

The nature of interaction between soil skeleton and pore water is determined by compressibility and shear resistance of soil skeleton, compressibility of pore water with gas inclusions and permeability of soil. The nature of interaction between soil skeleton and pore water is also influenced by the interaction of mineral particles and surrounding water molecules. The latter interaction is electrochemical in nature and occurs due to the presence of ions or atoms with excess electrons.

The solid and liquid phases participate in an ion exchange in a restricted layer known as the diffused layer. The thickness of this layer is determined by the quantity of loosely bonded or osmotically bonded water and also depends on the composition of the exchanged cations. As the thickness of the diffused layer in clayey soils affects their shear behaviour, any change in the composition of exchanged cations is capable of causing hardening or weakening of the clayey soil as a whole. It may thus be concluded that the rheological properties of clayey soil would depend on its mineralogical composition as well as the composition of the exchanged cations.

The interaction between solid particles in the soil skeleton is also complex in nature and is governed primarily by the mineralogical composition and the composition of the exchanged cations. Each particle of clayey soil interacts with surrounding particles either directly or through bound water. The net effect of these interactions is instrumental in producing the internal force field which is resisted by the external fields of gravitational and seepage forces. Thus, the internal forces of interaction in clayey soils play the same role as the forces of molecular interaction in solid, undispersed, crystalline media. As the forces of interaction between particles in clayey soil are many times less than the forces of molecular interaction between particles in the crystal lattice, the properties of clayey soils differ significantly from those of solid crystalline bodies.

The interaction of particles through bonded water in clayey soils promotes the formation of a coagulational bond which is mobile and is easily restored after disturbance. These bonds are known as coagulational thixotropic bonds and the bonds between the particles are referred to as primary bonds. The interaction between particles, whether directly or through cementing bonds, promotes the formation of a condensational-crystallisational bond and primary bonds. It may be mentioned that the condensational crystallisational bond is stronger than the coagulational thixotropic bond but is not restored after disturbance. The bonds of this type appear between particles as a result of physical, biochemical, chemical and other processes and are known as secondary bonds.

In recent publications the problem of interaction between particles is invariably treated with reference to coagulational bonds, in which the particles move closer or away under the effect of external forces. However, it is obvious that under the effect of an external force the particles will not only be attracted or repulsed, but will also become displaced and rotated with respect to one another.

For a coagulational bond the interaction between particles is dealt with for the simplest case, when they are located parallel to each other and are subjected to two external forces, normal force (N) and tangential force (T). It is evident that the internal forces of interaction between particles are stimulated by the external forces (Fig. 1.1a). As the forces of internal interaction between particles may have different signs, the external and the normal forces will affect the interaction in different ways. If the repulsive forces predominate, they will tend to balance the external normal forces. The tangential forces are balanced by viscoelastic forces of resistance of bonded water in the interparticle space. Consequently, the smaller the thickness of the bonded water film, the higher the aforesaid resistance. In general, when the particles are located randomly, the external forces are balanced by the resistance of particles to their approach (compaction) and mutual slip and rotation (distortion). This means that the coagulational bond has all the properties of a rigid body, i.e., it is distinguished

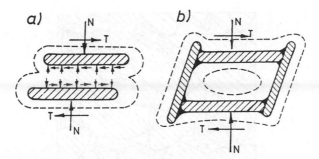

Fig. 1.1. Schematic representation of interaction between particles of clayey soils with (a) coagulational bond and (b) condensational thixotropic bond.

by its ability to resist volumetric changes as well as distortion. This conclusion is confirmed by the results of tests conducted on clays with disturbed structure (pastes).

In a condensational-crystallisational bond (Fig. 1.1b) the interaction between particles is manifested through rigid bonds which are broken during relative displacement or rotation between particles. All such bonds taken together impart spatial rigidity or structural strength to the soil. This structure can be disturbed by applying a certain force equal to the structural strength of soil. It is obvious that the structural strength under three-dimensional compression will be higher than under pure shear, i.e., the soil structure is more sensitive to shear.

Depending on the granulometric composition of clay and the degree of aggregation of particles, one or the other bond and, consequently, one or the other type of linkage between particles becomes predominant. The nature of interaction between particles in clayey soil is also significantly affected by the relative location of the bonds in space, i.e., by the structure. The formation of structure is a complex process and it depends on the physical and geographical conditions, accumulation of sediments, granulometric composition of particles etc.

In recent publications the structure of clayey soils is represented as follows: sand grains are surrounded by structured aggregates of clay and colloidal particles and exert very little effect on clay properties. However, when sand and clay particles are in a particular ratio in the soil, they may produce a bistructured skeleton which interacts with pore water in a complex manner (Fig. 1.2).

The interaction between soil skeleton and free water with dissolved gases also has a complex spatial-temporal character and is governed by the mechanical properties of the skeleton and pore water. As it is possible to estimate these properties individually, their interaction can also be quantitatively assessed by

8

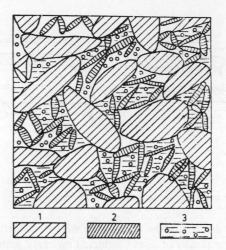

Fig. 1.2. Bistructured skeleton from sand and clay particles.

1—sand particles, 2—clay and colloidal particles, 3—free water with dissolved gas.

applying the principle of effective stresses, i.e., by using two systems of stresses and strains, one for the skeleton and the other for pore water.

3. Creep of Clayey Soils

The rheological properties of clayey soils manifest themselves differently under shear and during compaction. In the first case, the deformation due to a constant load may decay, remain steady or progressively increase, depending on the magnitude of shearing stresses (Fig. 1.3). If the working stress is less than the stipulated instantaneous strength but greater than the standard strength $(\tau_m > \tau > \tau_{mid})$, then the creep process ends in a progressive flow. If the working stress is less than the standard strength but greater than the long-term strength $(\tau_l^* < \tau < \tau_{mid})$, then the creep process does not end in steady flow at constant rate. Finally, if the working stress is less than the long-term strength $(\tau < \tau_l^*)$, then the creep process is seen to decay.

In the second case, the deformation due to a constant load always has a tendency to decay with time. However, the magnitude and rate of settlement may differ, depending on the degree of saturation and drainage conditions.

In clayey soils the magnitude and the rate of deformation during compression are significantly affected by the degree of water saturation of soil and the drainage conditions (Fig. 1.4). During intensive compression, the rate of deformation of water-saturated soil will be significantly affected by the coefficient of permeability if there is provision of drainage. The smaller the coeffi-

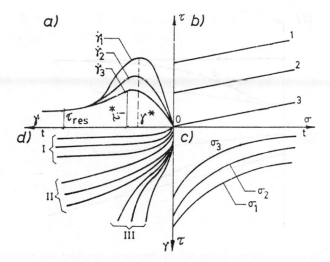

Fig. 1.3. Generalised rheological diagram of soil water under simple shear.

a—variation of shear resistance depending on rate of shear ($\dot{\gamma}_1 > \dot{\gamma}_2 > \dot{\gamma}_3$); *b*—limiting shear resistance at different shear rates (curves 1 and 2) and at $\gamma \gg \gamma^*$ (curve 3); *c*—long-term strength curves at various compacting loads; *d*—creep curves at different values of shearing stresses (group I—$\tau < \tau_{stab}$; group II—$\tau_{stab} > \tau > \tau_I^*$; group III—$\tau > \tau_I^*$).

cient of permeability, the lower will be the rate of settlement during consolidation.

The rheological properties of clayey soils are significantly dependent on the content of clayey fractions, water retention capacity of the minerals, density and moisture content of soil and the magnitude of the applied normal and shearing stresses. Considerable experimental data have been collected from studies devoted to the effect of the above-mentioned factors on the rheological properties [1, 3, 8, 13, 14]. Two of these factors, viz., density of dry soil and moisture content, will be detailed below as they are the deciding factors for a given granulometric and mineralogical composition.

The effect of the density of dry soil on its rheological properties can be easily established by testing specimens of different density, keeping the moisture content constant. This is achieved by precompression of soil under the maximum consolidating load, followed by application of shearing stresses at different levels of consolidating load. It may be mentioned that the traditional methods of testing of clayey soils on triaxial and other set-ups fail to account for the variation of soil density during the application of confining pressure and the subsequent deviatory loading. This results in large errors when the test results are processed. All the parameters determined from experimental curves imply the density to

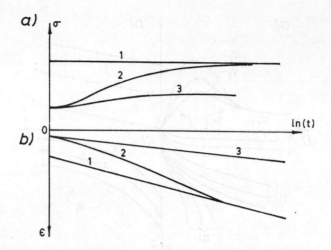

Fig. 1.4. (a) Variation of effective stresses and (b) creep curves for clayey soils under compression. 1—when degree of saturation $S_r < 0.8$; 2—when degree of saturation lies in the range $0.8 < S_r \leq 1$ and there is provision for drainage; 3—when degree of saturation lies in the range $0.8 < S_r < 1$ and there is no drainage.

be equal to the initial soil density, though actually it may experience significant variation. For instance, the curves of variations of the limiting shear resistance of a clayey soil of a given moisture content with normal stress differ significantly from each other, depending on the loading trajectory (Fig. 1.5). However, when the specimens were first consolidated and then subjected to loading by different trajectories, the experimental limiting shear resistance curves were found to be parallel to each other (Fig. 1.6).

From the analysis of the above test results and the data from other experimental studies on soil specimens of different initial densities, it was established that for various clayey soils at constant moisture content the angle of inclination of the limiting curve is independent of the density of dry soil. However, the ordinate of the starting point on the limiting curve is significantly dependent on the density of dry soil. Consequently, if the deformation is accompanied by change in density of dry soil, it can lead to hardening due to increase in cohesion. In view of the above, the change in cohesion must be taken into account while describing the strength criteria.

If the relation that governs the variation of cohesion with density or working stresses is known, then the condition of limiting equilibrium with consideration of hardening can be written without difficulty by assuming that the angle of internal friction is not affected much by the density of the dry soil.

$$\tau_\nu^* = \sigma_\nu \, \text{tg} \, \varphi + \sigma_\nu \, \text{tg} \, \alpha + c_0, \qquad \qquad \dots (1.2)$$

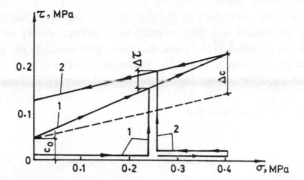

Fig. 1.5. Variation of shear resistance with loading trajectory for unsaturated clayey soil of a given moisture content.

1—loading trajectory without consolidation; 2—loading trajectory with consolidation under maximum load.

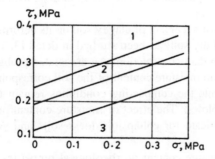

Fig. 1.6. Variation of limiting shear resistance with consolidating load, based on the results of tests conducted on a shearing set-up along a single plane after consolidation of specimens at moisture content $w = 0.12$.

1—at $\rho = 1.6$ g/cm^3; 2—at $\rho = 1.5$ g/cm^3; 3—at $\rho = 1.4$ g/cm^3.

where τ_ν^* is shear strength along the shear plane, σ_ν is the normal consolidating stress on the shear plane, φ is the true angle of internal friction, α is the angle due to hardening and c_0 is the cohesion corresponding to the initial density of dry soil ρ_0.

Equation (1.2) may be written in the form:

$$\tau_\nu^* = \sigma_\nu \, \mathrm{tg} \, \varphi' + c_0,$$

where φ' is the apparent angle of internal friction. If equation (1.2) is substituted in the Coulomb-Mohr* strength criterion $|\tau_\nu| = \tau_\nu^* = c_0 + \sigma_\nu \, \mathrm{tg}\, \varphi$ and an attempt is made to find the orientation of the slip field in the stress space, then it is found that in accordance with the Coulomb-Mohr strength theory the slip planes pass through the principal axes σ_2 and that these planes satisfy the condition of extremum of the function $f(\tau_\nu - \tau_\nu^*)$. Assuming $l = \cos\psi$, $n = \sqrt{1 - l^2}$ and $m = 0$ to be direction cosines for σ_ν and τ_ν and keeping in view that $\sigma_\nu = \sigma_3 l^2 + \sigma_2 m^2 + \sigma_1 n^2$; $\tau_\nu^2 = (\sigma_3 - \sigma_1)^2 l^2 m^2 + (\sigma_2 - \sigma_1)^2 m^2 n^2 + (\sigma_1 - \sigma_3)^2 l^2 n^2$; $l^2 + m^2 + n^2 = 1$, it is found that

$$|\tau_\nu| = \tfrac{1}{2}(\sigma_1 - \sigma_3)\sin 2\varphi,$$

$$\sigma_\nu = \frac{\sigma_1 + \sigma_3}{2} \, \frac{\sigma_1 - \sigma_3}{2} \cos 2\varphi.$$

The condition of extreme of the function $f(\tau_\nu - \tau_\nu^*)$ yields the following:

$$\mathrm{ctg}\, 2\psi = \mathrm{tg}\, \varphi' \text{ or } \psi = \pi/4 \pm \varphi'/2. \qquad \ldots (1.3)$$

It follows from the above that the slip fields are inclined at $45° \pm \varphi'/2$ to the axis σ_3 and that in the course of deformation, the orientation of the planes deviates from the initial position and tends to $45° \pm \varphi'/2$ with increase of stresses and hardening of the soil.

The effect of moisture content of clayey soil on its deformation and strength for a given density of dry soil has been studied in detail [3, 13] because clayey soils are often used in dam construction. An associated problem deals with determination of optimum moisture content of the soil corresponding to maximum compaction effect while the compacting conditions remain the same (number of loading cycles or blows). The effect of moisture content on deformation and strength is also significant for collapsible loess soils and swelling soils with undisturbed structure.

The effect of moisture content on rheological properties of clayey soils at constant density of dry soil can be attributed to the interaction between water and mineral particles that has a plasticising influence. Capillary tension forces appear with increase in moisture content but disappear as the state of total water saturation is approached.

From an analysis of results of tests on clayey soils at varying moisture content and constant density, it was observed that the deformation behaviour does not change significantly after reaching a certain moisture content, which is less than the moisture content corresponding to total water saturation [20]. This result indicates that there exists a particular moisture content limit beyond which the surface potential of mineral particles ceases to be effective, i.e. mineral particles are completely surrounded by diffused films of osmotically bonded water. In other words, such a state of moisture content corresponds to the soil

---

* *sic*; usually written Mohr-Coulomb—Technical Editor.

skeleton, which consists of mineral particles and bonded water. Obviously, if the moisture content of soil corresponding to the soil skeleton of a given mineralogical composition is known, then it suffices to test the soil specimens of this particular moisture content to derive the mechanical properties of the skeleton. However, the knowledge of such a relation is necessary only for unsaturated soils in which the density and moisture content change during the course of loading.

The problem of the effect of soil dampness on its mechanical properties is linked with the applicability of Terzaghi's principle of effective stresses. According to this principle the effective stresses acting on the soil skeleton are the cause of its deformation and failure. Therefore, to describe the stress-strain state of saturated clayey soil it is necessary to know the mechanical properties of the soil skeleton in addition to the magnitude of stresses in the skeleton. It therefore follows that by studying the behaviour of soil at a particular density and at different moisture contents, including full saturation, it is possible, firstly, to establish the magnitude of the moisture content of the soil skeleton and secondly, to check the applicability of Terzaghi's principle of effective stresses to clayey soils.

It may be mentioned in conclusion that the change of moisture content of unsaturated loess and swelling soils not only reduces their strength and deformability, but also changes the orientation of slip planes in the stress space, i.e., $\psi = f(w) = 45° \pm \varphi(w)/2$. This factor plays an important role in describing the plastic deformation of such soils in the stage just preceding limiting strength.

The foregoing discussion on the effect of density and moisture content on the strength of clayey soils is also fully valid for the effect of these factors on the rheological properties of clayey soils.

2

Stresses, Strains and Displacements in Clayey Soils

1. General Principles

Clayey soil, like any other soil, is a dispersive medium. However, it has a more complex structure due to the interaction between the skeleton and pore water with dissolved gases. Therefore, to describe the stress-strain state in such a medium, it is necessary to examine two stress systems, one for the skeleton and the other for pore water. As the shearing stresses in pore water are zero, only one stress remains. This is known as pore pressure and it is equal in all directions.

The stress and strain components in soil mass must conform to the initial and boundary conditions of the problem under consideration. It is likewise essential that the relation between stress and strain at an arbitrary point of the soil mass should be in conformity with the mechanical properties of the soil or the equations of state.

2. Fundamentals of Stresses, Strains and Displacements

The intensity of external or internal forces, i.e., force per unit area, is called stress. Stresses may act in directions normal and tangent to the surface of the area under consideration. The sides of an elementary parallelepiped isolated from the soil mass will experience normal and shearing stresses that tend to deform the parallelepiped, i.e, they tend to elongate the edges and change the angle between them. The ratio between the increment of length of the edge and the initial length of edge is known as linear strain, while the change in the angles between the edges is known as shearing strain (Fig. 2.1).

The stress-strain state in an elementary volume (point) of the soil can be established if we know six stress components (σ_x, σ_y, σ_z, $\tau_{xy} = \tau_{yx}$, $\tau_{xz} = \tau_{zx}$, $\tau_{yz} = \tau_{zy}$), six strain components (ε_x, ε_y, ε_z, $\gamma_{xy} = \gamma_{yx}$, $\gamma_{xz} = \gamma_{zx}$, $\gamma_{yz} = \gamma_{zy}$) and three components of displacements (u, v, w), interrelated through physical (governing) equations. The equality of shearing stresses follows from the equality of moments about the centre of the parallelepiped, i.e.,

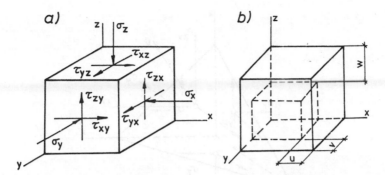

Fig. 2.1. (a) Components of stresses and (b) displacements in an elementary parallelepiped.

$\tau_{xy}d_xd_yd_z = \tau_{yx}d_xd_yd_z$. Similarly, the equality of shearing strains follows from the reciprocity of the shearing deformations of the parallelepiped.

In the discussion to follow, compressive stresses and linear compressive strain will be assumed as positive. Besides, the deformation of an elementary volume of soil will be taken as the deformation of the mineral soil skeleton. In a closed system the volumetric strain of pore water is assumed to depend on the volumetric strain of the skeleton, i.e., $\varepsilon_s = n\varepsilon_w$ (where n is soil porosity).

In engineering practice one often encounters problems in which it is necessary to determine the stress components in a plane that is arbitrarily inclined with respect to the x, y and z axes. Let the surface with normal ν be selected as such a plane (Fig. 2.2). Let l, m and n denote the cosines of the angles that normal ν make with the co-ordinate axes. Keeping in view that $l^2 + m^2 + n^2 = 1$, the following relations are obtained from the condition of equilibrium of the tetrahedron.

$$\left. \begin{array}{l} p_{x\nu} = \sigma_x l + \tau_{xy}m + \tau_{xz}n; \\ p_{y\nu} = \tau_{xy}l + \sigma_y m + \tau_{yz}n; \\ p_{z\nu} = \tau_{xz}l + \tau_{yz}m + \sigma_z n. \end{array} \right\} \qquad \ldots (2.1)$$

It is obvious that the total stress on the plane under consideration will be:

$$p_\nu = \sqrt{p_{x\nu}^2 + p_{y\nu}^2 + p_{z\nu}^2}.$$

By resolving the total stress along the normal and tangent to the surface, it is possible to determine the normal and shearing stresses on the given plane, i.e., $\sigma_\nu = p_{x\nu}l + p_{y\nu}m + p_{z\nu}n; \sigma_\nu^2 + \tau_\nu^2 = p_\nu^2$.

At a particular inclination of the plane the shearing stresses on the area will be zero, i.e., $\tau_\nu = 0$ and $\sigma_\nu = p_\nu$. At every point it is possible to identify at least three mutually perpendicular planes on which the shearing stresses are zero. These planes are known as principal planes. The stresses in these planes are known as principal stresses and they are denoted as σ_1, σ_2 and σ_3 such that

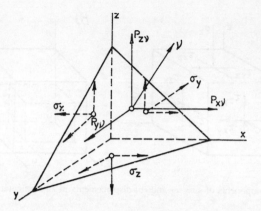

Fig. 2.2. Stress component on a plane, arbitrarily inclined with respect to x, y and z axes.

$\sigma_1 \geq \sigma_2 \geq \sigma_3$. In a similar manner it is possible to draw through every point of the body three mutually perpendicular planes on which the shearing stresses are maximum. These stresses are known as principal shearing stresses and they are related to the principal normal stresses by the following expressions:

$$\tau_1 = (\sigma_2 - \sigma_3)/2; \quad \tau_2 = (\sigma_1 - \sigma_3)/2; \quad \tau_3 = (\sigma_1 - \sigma_2)/2. \qquad \dots (2.2)$$

Hence, the principle normal and shearing stresses are invariant quantities which are independent of the selected direction of the co-ordinate axes.

The nine stress components mentioned above constitute the following stress tensor:

$$\left\{ \begin{matrix} \sigma_x & \tau_{yx} & \tau_{zx} \\ \tau_{xy} & \sigma_y & \tau_{zy} \\ \tau_{xz} & \tau_{yz} & \sigma_z \end{matrix} \right\}. \qquad \dots (2.3)$$

In general, the state of stress may be represented as a sum of two states, described by identical normal stresses σ in the co-ordinate planes (Fig. 2.3), normal stresses $s_x = \sigma_x - \sigma$, $s_y = \sigma_y - \sigma$, $s_z = \sigma_z - \sigma$ and shearing stresses $\tau_{xy} = \tau_{yx}$; $\tau_{yz} = \tau_{zy}$; $\tau_{zx} = \tau_{xz}$. Further, $3\sigma = \sigma_x + \sigma_y + \sigma_z$ and $s_x + s_y + s_z = 0$. Keeping the above in mind, the stress tensor may be written as follows:

$$T_\sigma + D_\sigma \left\{ \begin{matrix} \sigma & 0 & 0 \\ 0 & \sigma & 0 \\ 0 & 0 & \sigma \end{matrix} \right\} + \left\{ \begin{matrix} s_x & \tau_{yx} & \tau_{zx} \\ \tau_{xy} & s_y & \tau_{zy} \\ \tau_{xz} & \tau_{yz} & s_z \end{matrix} \right\}. \qquad \dots (2.4)$$

Tensor T_σ is known as the spherical tensor and represents all-round compression; tensor D_σ is known as the stress deviator and indicates how much the given state of stress differs from all-round compression.

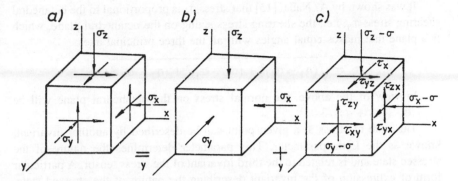

Fig. 2.3. (a) Resolution of stress tensor into (b)* spherical tensor and (c) stress deviator.

It can be shown that principal stresses are given by the real roots of the cubic equation given below:

$$\sigma^3 - I_{1\sigma}\sigma^2 - I_{2\sigma}\sigma - I_{3\sigma} = 0,$$

where

$$\left.\begin{array}{l} I_{1\sigma} = \sigma_x + \sigma_y + \sigma_z; \\ I_{2\sigma} = -\sigma_x\sigma_y - \sigma_y\sigma_z - \sigma_z\sigma_x + \tau_{xy}^2 + \tau_{yz}^2 + \tau_{zx}^2; \\ I_{3\sigma} = \sigma_x\sigma_y\sigma_z - \sigma_x\tau_{yz}^2 - \sigma_y\tau_{zx}^2 - \sigma_z\tau_{xy}^2 + 2\tau_{xy}\tau_{yz}\tau_{zx}. \end{array}\right\} \quad \dots (2.5)$$

As the principal stresses are invariant quantities that are independent of the selected co-ordinate axes, the coefficients of the cubic equation do not change with rotation of the co-ordinate axes, i.e., they are also invariants and are known as the first ($I_{1\sigma}$), second ($I_{2\sigma}$) and third ($I_{3\sigma}$) invariants of the stress tensor.

Assuming the shearing stresses to be zero, it is easy to obtain from eqn. (2.5) the expressions for the three invariants of the stress tensor in terms of principal stresses as follows:

$$\left.\begin{array}{l} I_{1\sigma} = \sigma_1 + \sigma_2 + \sigma_3; \\ I_{2\sigma} = -\sigma_1\sigma_2 - \sigma_2\sigma_3 - \sigma_3\sigma_1; \\ I_{3\sigma} = \sigma_1\sigma_2\sigma_3. \end{array}\right\} \quad \dots (2.6)$$

The invariants of the spherical tensor and stress deviator can be obtained by taking into account (2.4). However, it is more convenient to deal with parameters that are proportional to the square roots of the stress invariants. These parameters are principal normal (σ_i) and shearing (τ_i) stresses:

$$\left.\begin{array}{l} \sigma_i = \sqrt{(\sigma_1 - \sigma_2)^2 + (\sigma_2 - \sigma_3)^2 + (\sigma_3 - \sigma_1)^2}/\sqrt{2}; \\ \tau_i = \sqrt{(\sigma_1 - \sigma_2)^2 + (\sigma_2 - \sigma_3)^2 + (\sigma_3 - \sigma_1)^2}/\sqrt{6}, \end{array}\right\} \quad \dots (2.7)$$

---

*σ_z, σ_y and σ_x in figure should read σ only—Technical Editor.

It is evident from the above expressions that $\sigma_i = \sqrt{3}\tau_i$.

It was shown by A. Nadai [15] that stress σ_i is proportional to the octahedral shearing stress τ_0, i.e., the shearing stress acting on the octahedral plane, which is a plane that makes equal angles with all the three principal axes:

$$\tau_0 = \sqrt{(\sigma_1 - \sigma_2)^2 + (\sigma_1 - \sigma_3)^2 + (\sigma_2 - \sigma_3)^2}/3.$$

In view of the above, the normal stress on the octahedral plane will be $\sigma = (\sigma_1 + \sigma_2 + \sigma_3)/3$.

The state of stress at a given point can be described by another invariant, known as the Lode parameter. This parameter determines the nature of the stressed state and is related to the third invariant of the stress tensor. A particular form of expression of the invariant describing the nature of the stressed state, known as the Nadai-Lode parameter, is most widely used:

$$\mu_\sigma = (2\sigma_2 - \sigma_1 - \sigma_3)/(\sigma_1 - \sigma_3), \qquad \ldots (2.8)$$

where $\sigma_1 \geq \sigma_2 \geq \sigma_3$.

Basically, this parameter is related with the angle ω between stress σ_i and the projection of axis σ_3 on the octahedral plane, i.e., σ_3' (Fig. 2.4). Consequently,

$$\mu_\sigma = \sqrt{3}\,\mathrm{ctg}[\omega + (\pi/3)]. \qquad \ldots (2.9)$$

Parameters μ_σ and ω vary over the range $-1 \leq \mu_\sigma \leq +1$ and $\pi/3 \geq \omega \geq 0$ respectively. For example, under three-dimensional axisymmetrical compression when $\sigma_1 > \sigma_2 = \sigma_3 > 0$, $\mu_\sigma = -1$ and $\omega = \pi/3$.

Let us examine in greater detail the deformation of the elementary parallelepiped. The position of the elementary parallelepiped in a deformed state can be established provided we know three displacement components u, v, and w, three strains $\varepsilon_x, \varepsilon_y$, and ε_z and three pairs of shear strain $\gamma_{xy} = \gamma_{yx}$; $\gamma_{yz} = \gamma_{zy}$ and $\gamma_{xz} = \gamma_{zx}$.

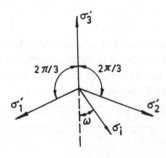

Fig. 2.4. Angle between stress σ_i and the projection of axis σ_3 on the octahedral plane.

Henceforth, it will be assumed that the strains are small quantities and hence their squares and products may be neglected. With this assumption, the following linear relations are established between strains and displacements:

$$\left.\begin{array}{c} \varepsilon_x = \dfrac{\partial u}{\partial x}; \quad \varepsilon_y = \dfrac{\partial v}{\partial y}; \quad \varepsilon_z = \dfrac{\partial w}{\partial z}; \\[2mm] \gamma_{xy} = \dfrac{\partial u}{\partial y} + \dfrac{\partial v}{\partial x}; \quad \gamma_{yz} = \dfrac{\partial v}{\partial z} + \dfrac{\partial w}{\partial y}; \\[2mm] \gamma_{xz} = \dfrac{\partial w}{\partial x} + \dfrac{\partial v}{\partial z}. \end{array}\right\} \qquad \dots (2.10)$$

According to the theory of elasticity, when the co-ordinate axes are rotated, the change in strain components is compatible with the change in the stress components. On comparing the two changes, it is found that the change in strain components may be obtained from the change in stress components by replacing σ and τ with ε and $\gamma/2$ with the appropriate subscripts. Consequently, all the conclusions of the theory of stress pertaining to the rotation of axes can be applied to the theory of strain. Hence, the six strain components, viz. ε_x, ε_y, ε_z, $\gamma_{xy}/2$, $\gamma_{yz}/2$ and $\gamma_{zx}/2$ will constitute the following strain tensor:

$$T_\varepsilon = \left\{ \begin{array}{ccc} \varepsilon_x & \gamma_{xy}/2 & \gamma_{xz}/2 \\ \gamma_{yx}/2 & \varepsilon_y & \gamma_{yz}/2 \\ \gamma_{zx}/2 & \gamma_{zy}/2 & \varepsilon_z \end{array} \right\}. \qquad \dots (2.11)$$

The above tensor can be represented as a superposition of two strains, one due to volumetric changes and the other due to distortion. The first one is described by identical strains ε, while the second is described by linear strains $e_x = \varepsilon_x - \varepsilon$, $e_y = \varepsilon_y - \varepsilon$, $e_z = \varepsilon_z - \varepsilon$ and shear strains γ_{xy}, γ_{yz}, γ_{zx}. It is obvious that the first strain is characterised by an absence of distortion and is described by the relation $3\varepsilon = \varepsilon_x + \varepsilon_y + \varepsilon_z$, while the second strain is characterised by absence of any volumetric change since $e_x + e_y + e_z = 0$.

The distinction between two types of strains has a certain physical significance, because the soil medium behaves differently while resisting changes in volume and shape. This applies, first of all, to plastic strain and plays an important role in predicting the consolidation processes that take place due to volumetric strain.

The first strain is described by the spherical strain tensor:

$$T_{\varepsilon_0} = \left\{ \begin{array}{ccc} \varepsilon & 0 & 0 \\ 0 & \varepsilon & 0 \\ 0 & 0 & \varepsilon \end{array} \right\}; \qquad \dots (2.12)$$

and the second strain by the strain deviator

$$D_\varepsilon = \left\{ \begin{array}{ccc} e_x & \gamma_{yx}/2 & \gamma_{zx}/2 \\ \gamma_{xy}/2 & e_y & \gamma_{zy}/2 \\ \gamma_{xz}/2 & \gamma_{yz}/2 & e_z \end{array} \right\} \qquad \dots (2.13)$$

It may be noted that $T_\varepsilon = T_{\varepsilon_0} + D_\varepsilon$.

Thus, the strain deviator indicates how much the deformed state under consideration deviates from that of all-round compression.

At every point of the deformed body there are three mutually perpendicular axes for which the shear strain components are zero. These axes are known as principal axes. The strains of these axes are known as principal strains and are denoted as ε_1, and ε_2 and ε_3. Principal strains are real roots of the following cubic equation:

$$\varepsilon^3 - I_{1\varepsilon}\varepsilon^2 - I_{2\varepsilon}\varepsilon - I_{3\varepsilon} = 0. \qquad \dots (2.14)$$

As the principal strains are invariant quantities, the coefficients of the cubic equation do not change with rotation of the co-ordinate axes, i.e., they are invariants of the strain tensor and are determined from the following expressions:

$$\left. \begin{array}{l} I_{1\varepsilon} = \varepsilon_x + \varepsilon_y + \varepsilon_z; \\ I_{2\varepsilon} = -\varepsilon_x\varepsilon_y - \varepsilon_y\varepsilon_z - \varepsilon_z\varepsilon_x + \gamma_{xy}^2/4 + \gamma_{yz}^2/4 + \gamma_{zx}^2/4; \\ I_{3\varepsilon} = \varepsilon_x\varepsilon_y\varepsilon_z - (\varepsilon_x\gamma_{yz}^2 + \varepsilon_y\gamma_{zx}^2 + \varepsilon_z\gamma_{xy}^2 - \gamma_{xy}\gamma_{yz}\gamma_{zx})/4. \end{array} \right\} \qquad \dots (2.15)$$

The relations for invariants of the strain tensor can be simplified by expressing them through principal strains:

$$I_{1\varepsilon} = \varepsilon_1 + \varepsilon_2 + \varepsilon_3;$$
$$I_{2\varepsilon} = -\varepsilon_1\varepsilon_2 - \varepsilon_2\varepsilon_3 - \varepsilon_3\varepsilon_1;$$
$$I_{3\varepsilon} = \varepsilon_1\varepsilon_2\varepsilon_3.$$

Extending the analogy between the states of stress and strain, the Nadai-Lode parameter for strain can be written as follows:

$$\mu_\varepsilon = (2\varepsilon_2 - \varepsilon_1 - \varepsilon_3)/(\varepsilon_1 - \varepsilon_3), \qquad \dots (2.16)$$

where $\varepsilon_1 \geq \varepsilon_2 \geq \varepsilon_3$.

The quantity proportional to the square root of the second invariant of the strain tensor is defined as strain. Depending on the coefficient of proportionality, strain is differentiated as linear and shearing. Linear strain is given by the expression:

$$\varepsilon_i = \frac{\sqrt{2}}{3}\sqrt{(\varepsilon_x - \varepsilon_y)^2 + (\varepsilon_y - \varepsilon_z)^2 + (\varepsilon_z - \varepsilon_x)^2 + \frac{3}{2}(\gamma_{xy}^2 + \gamma_{yz}^2 + \gamma_{zx}^2)}$$
$$\dots (2.17)$$

while shearing strain is described by the expression:

$$\gamma_i = \sqrt{\frac{2}{3}} \sqrt{(\varepsilon_x - \varepsilon_y)^2 + (\varepsilon_y - \varepsilon_z)^2 + (\varepsilon_z - \varepsilon_x)^2 + \frac{3}{2}(\gamma_{xy}^2 + \gamma_{yz}^2 + \gamma_{zx}^2)}.$$
$$\dots (2.18)$$

Comparing the two expressions, it is found that $\gamma_i = \sqrt{3}\varepsilon_i$. Therefore, linear strain and shearing strain may be expressed through principal strains as follows:

$$\varepsilon_i = \frac{\sqrt{2}}{3} \sqrt{(\varepsilon_1 - \varepsilon_2)^2 + (\varepsilon_2 - \varepsilon_3)^2 + (\varepsilon_1 - \varepsilon_3)^2};$$

$$\gamma_i = \sqrt{\frac{2}{3}} \sqrt{(\varepsilon_1 - \varepsilon_2)^2 + (\varepsilon_2 - \varepsilon_3)^2 + (\varepsilon_1 - \varepsilon_3)^2}.$$

The octahedral shearing strain may also be expressed through principal strains:

$$\gamma_0 = \frac{2}{3} \sqrt{(\varepsilon_1 - \varepsilon_2)^2 + (\varepsilon_2 - \varepsilon_3)^2 + (\varepsilon_1 - \varepsilon_3)^2}.$$

For the case of pure shear $\varepsilon_x = \varepsilon_y = \varepsilon_z = 0$; $\gamma_{xy} = \gamma$; $\gamma_{yz} = \gamma_{zx} = 0$; $\gamma_i = \gamma$; $\varepsilon_i = \gamma/\sqrt{3}$; $\gamma_0 = \sqrt{2}\varepsilon_i = \sqrt{2}\gamma/\sqrt{3}$; for the case of constrained uniaxial compression $\varepsilon_x = \varepsilon_y = 0$; $\varepsilon_v = \varepsilon_z$; $\varepsilon_i = 2\varepsilon_z/3$; $\varepsilon = \varepsilon_z/3$.

The notations of stress and strain components adopted in the present work are standard notations, particularly in engineering practice. However, for the sake of brevity it is sometimes more convenient to use the tensor symbols. In this case, the nine components of stress and strain can be represented by a single symbol each, i.e., σ_{ij} and ε_{ij}, where i and j are equal to 1, 2, 3. When $i = j$, the symbols σ_{11}, σ_{22}, σ_{33} represent normal stresses and the symbols ε_{11}, ε_{22}, ε_{33}, represent normal strains. When $i \neq j$, τ_{12}, τ_{23}, τ_{31} represent shearing stress components, while ε_{12}, ε_{23}, ε_{31} represent shearing strain components.

Pore pressure and effective stress

The concepts of stress and strain were defined above with application to the soil medium as a whole. However, these concepts are valid for multiphase soils only when the process of formation of stress and strain fields is complete, i.e., when it has stabilised.

At present, the Terzaghi principal of effective stresses is employed to describe the unstable stress-strain state of multiphase soil. According to this principal, the total stress in multiphase soil may be represented as a sum, consisting of stress in the soil skeleton and pore pressure in the pore water. It is assumed that only stresses acting in the soil skeleton can produce its consolidation and increase its shear strength. These stresses are known as effective stresses. Such a distinction between stresses obviously assumes a mechanical model for soil consisting of skeleton and pore water. Hence it is necessary to first establish what is meant by the term 'soil skeleton'.

Soil skeleton is understood to be an aggregate of mineral particles which together with bound water constitutes a complex structure (frame) that is capable of resisting changes in volume and shape. The mass of bound water attached to the soil skeleton varies, depending on the mineral and granulometric composition of the soil. However, it is independent of the density of mineral particles in the soil. It is extremely difficult to determine the moisture content of such a skeleton, which explains the almost total absence of experimental studies on this topic. Since the mechanical properties of the soil skeleton are not supposed to change when the moisture content exceeds that of the soil skeleton, it is possible to determine the moisture content of the skeleton by testing identical soil specimens of a given density with different moisture contents.

It may be mentioned that such a mechanical model does not always describe the actual behaviour of the soil under load. For instance, the high-dispersion montmorillonite clays retain the coagulational structure over a wide range of moisture content. For such a structure, the concept of skeleton is not associated with direct interaction between particles. In this case, the structure is formed due to the internal force field, i.e., due to surface energy of particles and osmotic forces in bond water. In such a model, there is no need for differentiation of stresses into effective stress and pore pressure [9], which pressure appears in such a system only when it exceeds the osmotic pressure acting in the interparticle space. The higher the density of such a soil, the greater the osmotic pressure and, consequently, the higher must be the applied load in order to produce excess pore pressure. To date there has been almost no direct experimental confirmation of the above model of clayey soils, known as the osmotic model. The author believes that this model is applicable only to clays with coagulational structure, including montmorillonite clays.

Notwithstanding the limitations discussed above, the principle of effective stresses has been confirmed by numerous experimental studies for many types of clayey soils, particularly loam and sandy soils which are distinguished by a predominantly rigid skeletal bond. Therefore, the soil model based on the principle of effective stresses will be adopted in all further discussion, i.e., soil will be assumed to consist of a load-bearing particle frame and pore water, both actively interacting in the process of deformation of soil mass. Interaction between the soil skeleton and pore water leads to a continuous redistribution of total stresses between the skeleton and pore water.

The effective stress and pore pressure provide a quantitative estimate of the interaction between the skeleton and pore water. It may be mentioned that these parameters are relevant for the unstabilised state of multiphase soil mass, during which there is intense variation of phase proportions in a unit mass of soil and intensive interaction between the phases. When the soil mass has stabilised, its state of stress can be described by a single stress system, because there is no interaction between the phases and external load is completely balanced by stresses in the skeleton.

During the deformation of soil mass, the effective stresses in the skeleton and the pore pressure in the liquid balance the effect of external load and dead weight. It is a well-known fact that such a state ends in total transfer of the external load to the skeleton, resulting in stabilisation of the soil mass.

In the unstabilised state, the total stress in multiphase soil is equal to the sum of effective stress and pore pressure, i.e.,

$$\sigma = \sigma' + u_w. \qquad \ldots (2.19)$$

A similar relation exists between principal stresses and pore pressure, i.e., $\sigma_v = \sigma'_v + 3u_w$ (where σ_v and σ'_v are the sum of total principal stresses and effective stresses respectively).

In conclusion, it may be mentioned that in multiphase soil, strain implies strain of the skeleton (frame), which is identically equal to the strain of soil as a whole, because the skeleton and soil occupy the same volume. As regards pore water, it offers resistance only to volumetric strain, which is related to volumetric strain of the skeleton of a closed soil system as follows: $\varepsilon = n\varepsilon_w$ (n is soil porosity and ε_w is average strain of pore water).

Equations of equilibrium and continuity

The stresses, strains, displacements and pore pressure and their rates of variation are related to each other through equations of equilibrium and compatibility of strain.

The equations of equilibrium of a single-phase medium may be obtained by examining the equilibrium of an elementary parallelepiped:

$$\left. \begin{array}{c} \dfrac{\partial \sigma_x}{\partial x} + \dfrac{\partial \tau_{xy}}{\partial y} + \dfrac{\partial \tau_{xz}}{\partial z} + \rho(F_x - a_x) = 0; \quad (x, y, z); \\ \ldots\ldots\ldots\ldots\ldots\ldots\ldots\ldots\ldots\ldots\ldots\ldots\ldots\ldots\ldots\ldots\ldots\ldots\ldots\ldots \end{array} \right\} \qquad \ldots (2.20)[1]$$

where F_x, F_y and Fz are body force components, a_x, a_y and a_z are acceleration components and ρ is density of the medium.

In tensor notation, these equations may be written in the form:

$$\dfrac{\partial \sigma_{ij}}{\partial x_i} + \rho(F_j - a_j) = 0,$$

where a_j represents acceleration components; i and $j = 1, 2, 3$.

For multiphase soil, the equilibrium equations are written in terms of effective stresses by substituting $\sigma'_x = \sigma_x - u_w$ in place of σ_x in (2.20).

---

[1] The remaining formulae may be obtained by changing the subscripts in cyclic order.

When the body is in a state of quasistatic equilibrium, i.e., when the accelerations of the particles are small and close to zero, the equations of equilibrium can be simplified and written as follows:

$$\left. \frac{\partial \sigma'_x}{\partial x} + \frac{\partial \tau_{xy}}{\partial y} + \frac{\partial \tau_{xz}}{\partial z} + X = 0 \ (x,y,z); \right\}$$

where σ'_x, σ'_y and σ'_z are components of effective stresses and x, y and z are components of volumetric forces.

The equations of continuity or compatibility of strain are derived from eqns. (2.10) by eliminating from them the components of displacements u, v and w, i.e.,

$$\left. \begin{array}{l} 2\dfrac{\partial^2 \varepsilon_x}{\partial y \partial z} = \dfrac{\partial}{\partial x}\left(-\dfrac{\partial \gamma_{zy}}{\partial x} + \dfrac{\partial \gamma_{zx}}{\partial y} + \dfrac{\partial \gamma_{x,y}}{\partial z} \right) \ (x,y,z); \\[4mm] \dfrac{\partial^2 \gamma_{xy}}{\partial x \partial y} = \dfrac{\partial^2 \varepsilon_x}{\partial y^2} + \dfrac{\partial^2 \varepsilon_y}{\partial x^2} \ (x,y,z); \end{array} \right\} \quad \ldots (2.21)$$

These six equations correlate the strain components in such a manner that when deformation takes place there is no discontinuity at any point of the medium, i.e., the condition of continuity of strain hold good for the body, thereby ensuring conformity with one of the fundamental tenets of continuum mechanics.

It may be mentioned that the same relations hold good for determining strain rates, except that the strain components should be replaced by the corresponding strain rate components ($\dot{\varepsilon}_x$, $\dot{\gamma}_{xy}$ etc.).

The equations of continuity for pore water during deformation and flow through a porous medium may be obtained from the condition of constancy of liquid mass and mineral particles in an elementary volume of soil [17]. The simplified equation of continuity may be written in the following form:

$$\frac{\partial \varepsilon}{\partial t} + n\frac{\partial \varepsilon_w}{\partial t} = \frac{k}{\gamma_w}\nabla^2 u_w, \qquad \ldots (2.22)$$

where ε and ε_w are volumetric strain of soil and pore water respectively; k is coefficient, of permeability; γ_w is unit weight of pore water and ∇ is the Laplace operator.

The left-hand side of eqn. (2.22) represents the volumetric changes in the soil skeleton and pore water of an elementary parallelepiped, while the right-hand side represents the volume of water flowing out of the parallelepiped or into it.

Equation (2.22) is known as the consolidation equation. It is widely used in soil mechanics and is valid for all manner of volumetric changes of the skeleton.

In the absence of seepage, i.e., when $k = 0$, eqn. (2.22) yields the condition of compatibility of strains of the skeleton and pore water, i.e., $\varepsilon = n\varepsilon_w$. For the case of incompressibility of skeleton in pore water, the well-known equation of steady-state seepage $\nabla^2 u_w = 0$ is obtained.

Initial and boundary conditions

Initial conditions are one of the basic conditions in solving applied problems of geomechanics and particularly the mechanics of multiphase soils, because they largely determine the selection of the analytical model and the methods of solution of boundary value problems. This applies first of all to the initial state of stress of the soil mass under consideration because the process of formation of additional stress fields depends on the stress level from which further loading of a point of soil mass is commenced.

Unfortunately, the initial state of stress is not known and it is impossible to specify it; determination of the initial state of stress is in itself a complex problem which cannot be solved without overcoming serious difficulties. The initial state of stress of the soil mass under consideration has formed over a long period and depends on a large number of factors, including homogeneity, isotropy, density and deformability of the soils making up the mass, history of formation of the soil mass, which is difficult to establish and simulate and, finally, the geometrical parameters of the soil mass (thickness of the layers, their inclination to the horizontal, dimensions, etc.). In view of these aspects, the initial state of stress of soil mass is usually specified on the basis of the simplest assumptions, which often results in large errors. In the present work, the author will later return to the question of initial state of stress of soil mass and the methods by which it may be measured and determined. Furthermore, it will be shown how this question is related to the selection of the analytical model of soil.

In publications on the theory of consolidation of soils, the initial state of stress of a multiphase soil mass under external load is determined by assuming that soil is free of volumetric strain, which is valid for fully saturated soils. In the method of reduced parameters of multiphase soil proposed by the author [17], it is not necessary to impose any constraints on the volumetric compressibility of soil. The only assumption made in [17] is that at the commencement of loading on the soil mass, there is no change in the relative proportion of the masses of the solid and liquid phase components. This assumption is valid for most clayey soils because they have a small coefficient of permeability and, therefore, during loading of the soil mass there is not enough time for expulsion of the pore water.

Hence the equations of equilibrium and continuity and the initial and boundary conditions for multiphase soil medium differ significantly from the traditional equations and conditions of the mechanics of a single-phase medium, developed with particular reference to structural materials.

The boundary conditions on the contour of soil mass may be specified in terms of stresses, displacements or heads as well as the rates of change of the above parameters. Mixed boundary conditions are also possible, when loads, displacements and pore pressure are partially specified on the boundary of the soil mass.

If the stresses on the boundary of the soil mass are given, then equilibrium condition (2.1) for an elementary tetrahedron (see Fig. 2.2) should be used, assuming that $p_{x\nu}$, $p_{y\nu}$ and $p_{z\nu}$ are boundary stresses, σ_x, σ_y, σ_z, τ_{yz}, τ_{xy} and τ_{xz} are internal stresses at points of the soil mass adjacent to the boundary and l, m and n are cosines of angles between normal ν to the contour at the given point and the co-ordinate axes.

It can be concluded from the foregoing that unlike the problems of continuum mechanics, it is necessary to write an additional boundary condition for pore pressure in order to solve boundary value problems of the mechanics of multiphase soils.

Types of strain in multiphase soil

Depending on the magnitude and duration of applied load, multiphase soil may experience elastic (reversible) and plastic (irreversible) deformation.

Elastic deformation occurs mostly under short duration and cyclic loads. Hence it is best to determine the characteristics of elastic deformation from the unloading curve obtained by conducting static tests. The elastic strain characteristics of soil K^e and G^e depend mainly on its density and moisture content and are not affected much by the state of stress of the soil. Thus the task of determining plastic strain and its parameters can be simplified to a large extent by separating elastic strain from total strain.

Plastic strain appears in soil both when it is in a prelimit as well as in a limit state of stress under loading. This considerably complicates the task of writing the governing equations. The difficulty in describing the plastic strain in the prelimit state arises due to the fact that it depends on a number of factors, viz., loading parameters, including the loading trajectory, type of stressed state and, finally, the initial state of stress and the initial density and moisture content of soil. Plastic (irreversible) strain develops in soil under all-round compression and especially under deviatory loading. It is noteworthy that in the latter case, the process of plastic deformation of soil is often accompanied by its dilatation, i.e., volumetric changes.

The features of mechanical properties of soils should be taken into account while writing the governing equations by various methods, based on the modern theories of plasticity, which are modified, however, to include the effect of volumetric plastic strain.

A number of significant achievements have been reported in this field in recent years, albeit sometimes contradictory.

It is obvious that the contradictions can be resolved only by conducting extensive numerical studies which would enable the measurement of volumetric and shearing strains with a high degree of accuracy.

It is rather difficult to consider the characteristics of elastic and plastic strain of soil in the governing equations because the range of variation of soil properties depends on its density, moisture content and granulometric and mineral composition etc. Any attempt at developing an equation of state that is adequate for all types of soils is bound to produce errors since all deformation theories are based on a number of assumptions and hypotheses that are valid only within certain limits.

Hence it is advisable to examine three stages of soil deformation for practical expediency. The first stage describes the linear relation between stress and strain, the second stage describes non-linear behaviour and the third stage pertains to creep (the third stage is absent under conditions of volumetric strain).

3. Rheological Equations of State

At present, various methods are employed for describing the rheological properties of clayey soils. These methods are based on mechanical models, theory of hereditary creep, engineering theory of creep, theory of plastic flow and molecular theory of flow. All the above methods have been described in detail by S.S. Vyalov [1]. Ultimately, all the above theories lead to different equations of state containing some parameters which need to be determined experimentally, i.e., the equations are phenomenological.

Mechanical models

The mechanical models represent the rheological properties of the soil skeleton by a combination of elastic, viscous and plastic elements. For example, the parallel combination of an elastic and a plastic element (Fig. 2.5a) lead to the Kelvin-Voight model and the following equation of state containing two parameters:

$$\tau = G_\infty \gamma + \eta \dot{\gamma}, \qquad \qquad \ldots (2.23)$$

where G_∞ is shearing modulus and η is viscosity.

For $\tau = $ const., the following equation is obtained by integrating (2.23):

$$\gamma(t) = \frac{\tau}{G_\infty} \left[1 - \exp\left(-\frac{G_\infty}{\eta} t \right) \right].$$

28

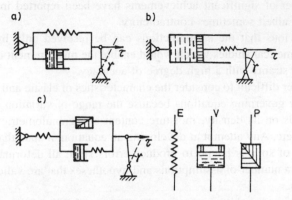

Fig. 2.5. Mechanical models representing rheological properties of soil.

a) Kelvin-Voight model; b) Maxwell model; c) Shvedov model; E—elastic element; V—viscous element; P—plastic element.

The above equation describes decaying creep. The quantity $n/G_\infty = T_{rel}$ is known as retardation time. However, it does not describe the relaxation process because at $\gamma = $ const. the stresses also remain constant, i.e., $\tau = G_\infty \gamma = $ const.

When the elastic and viscous elements are joined in parallel (Fig. 2.5b), the Maxwell model is obtained, for which the equation of state is as follows:

$$\tau = T_{rel}\dot{\tau} + \eta\dot{\gamma}, \qquad \ldots (2.24)$$

where $T_{rel} = \eta/G_0$ is relaxation period (here G_0 is the initial, stipulated instantaneous shearing modulus).

Equation (2.24) represents non-decaying creep and relaxation: at $\tau = $ const. it becomes the Newton equation $\dot{\gamma} = \tau/\eta$, while at $\eta = $ const., it becomes the relaxation equation $\tau = \tau_0 \exp(-t/T_{rel})$ (where τ_0 is the initial stress).

When the elastic element is joined in series with the viscous element and then in parallel with the Saint Venant plastic element (Fig. 2.5c), the Shvedov model, described by the following equation of state, is obtained:

$$\dot{\gamma} = (\tau - \tau^*)/\eta + \tau/G, \qquad \ldots (2.25)$$

where τ^* is the ultimate strength of the plastic element.

It is obvious that there is no creep deformation at $\tau < \tau^*$.

In applying the principals of mechanical modeling, several attempts were made to develop complex rheological models for reliable description of soil properties. Some researchers who have developed mechanical models are N.M. Maslov, S.S. Vyalov, M.N. Goldstein, Yu.K. Zaretskii, Muroyama, Shibati etc. For example, M.N. Maslov extended Shvedov's model and proposed a rheological model in which the ultimate strength of the plastic element is determined from the Coulomb-Mohr strength condition, i.e., $\tau^* = \sigma \,\mathrm{tg}\,\varphi + c$. Because of this feature, Maslou's model has found wide application in engineering practice.

Analysis of the mechanical models of soil has shown that in most cases they describe the rheological properties of soil under shear. This can be attributed to the history of development of the science of rheology of soils because the initial thrust of this science was on prediction of shear strain. Some of the above models may also be used for describing the strain under consolidation. For this purpose the equations of state must be modified by substituting σ, E_k and η_v in place of τ, G and η respectively (here σ is consolidating load, E_k is modulus of linear compression and η_v is viscosity under volumetric compression).

Different elements were placed in a cylinder with a piston (Fig. 2.6) to describe the process of consolidation of water-saturated clayey soils by means of mechanical models. However, all these models have one drawback: they describe the mechanism of transmission of load to the skeleton and water rather than the process of soil consolidation. The latter is associated not only with the redistributing of stresses between skeleton and pore water, but also with expulsion of water from the pores throughout the height of the consolidating soil layer. Consequently, these models are deficient in one geometrical parameter—the length of the drainage path. Yet, this very parameter plays the deciding role in the consolidation process. Given the foregoing, it would have been more appropriate to simulate the consolidation process for water-saturated clayey soils by means of skeleton models joined in sequence and placed in a cylinder of given height. In this case, the delay in deformation would occur not as much due to creep of soil skeleton, as due to length of drainage path (h). It is known that for a skeleton with elastic properties the period of stabilisation of the consolidation process is proportional to the square of the length of drainage path or the thickness of consolidating layer, i.e., $t_1/t_2 = (h_1/h_2)^2$.

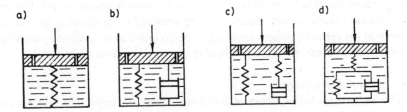

Fig. 2.6. Mechanical models representing the process of soil consolidation.

a) Terzaghi-Gersevanov model; b) Taylor's model; c) Tan's model; d) Gibson-Lo model.

On the basis of the models depicted in Fig. 2.7, it is possible to derive the equation for unidimensional consolidation with initial and boundary conditions. However, this is not possible with the models of Fig. 2.6. On the basis of results of tests on clayey soils, the author developed a mechanical model for representing and describing the process of consolidation in clayey soils (Fig. 2.7c). Unlike other models, in this model the main cylinder is filled with compressible fluid which simulates pore water with dissolved air bubbles.

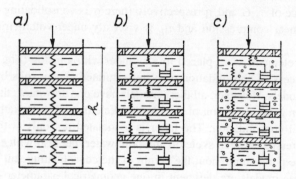

Fig. 2.7. Mechanical models representing the process of consolidation in a soil layer of thickness *h*.

a) with elastic skeleton and incompressible pore water; b) with elastoviscous skeleton and incompressible pore water; c) with elastoviscous skeleton and compressible pore water.

In conclusion, let us examine a few relatively complex mechanical models that are capable of simultaneously representing both the shear and volumetric creep of the soil skeleton.

Such a spatial model can be built using hinged connections between the basic linear elements inclined towards each other (Fig. 2.8). In this case, it is possible to describe the most diverse properties of the soil skeleton, depending on the relative position of the basic models in the overall spatial model. For instance, elastic volumetric deformation can be obtained by positioning the elastic elements in the central portion. Anisotropic behaviour can be studied by varying the properties of elements in perpendicular directions and so on.

Theory of hereditary creep

The Boltzmann-Volter theory of hereditary creep was applied to soils for the first time by V.A. Florin who modified the basic approach developed by Maslov and Arutyunyan. The validity of the theory was experimentally confirmed by S.P. Meschyan. According to this theory, the total shearing or volumetric strain

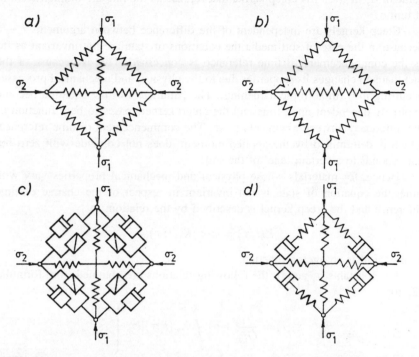

Fig. 2.8. Plane and spatial mechanical models representing the deformation behaviour of soil under simultaneous change of volume and shape.

a) elastic isotropic medium; b) anisotropic medium; c) viscoelastic medium,* d) elasto-viscoplastic medium.*

under arbitrary loading can be written in the form:

$$
\left.
\begin{aligned}
\varepsilon_i &= \frac{\sigma_i(t)}{G_m} + \int_{\tau_1}^{t} K_\gamma(t,\tau)\sigma_i(\tau)d\tau; \\
\varepsilon_v &= \frac{\sigma(t)}{K_m} + \int_{\tau_1}^{t} K_v(t,\tau)\sigma(\tau)d\tau,
\end{aligned}
\right\}
\quad \ldots (2.26)
$$

where G_m and K_m are instantaneous shear modulus and bulk modulus respectively, $K_\gamma(t,\tau)$ is creep kernel that represents the rate of shear strain for unit

---

*Transposed in Russian original; "c" should be "d" and vice versa—Technical Editor.

value of σ_i; $K_v(t, \tau)$ is creep kernel that represents the rate of volumetric strain at unit σ.

Creep kernels are independent of the difference between arguments $t - \tau$ because in the case of soil media the equations of state are not invariant as far as the commencement of time reference is concerned. This is because of the fact that soil changes its properties due to the physical and mechanical processes occurring in it (thixotropic hardening). The parameters of the equation become explicitly dependent upon time and the creep kernel ceases to be a function of the difference between arguments $t - \tau$. The commencement of time reference, which is determined by the applied moment, does not coincide with zero but corresponds to a certain 'age' of the soil.

Hence, for materials whose physical and mechanical properties vary with time, the equation of state is not invariant in respect of the change of time reference and the creep kernel is described by the relation:

$$K(t, \tau) = \varphi(\tau) K_1(t, \tau),$$

where $\varphi(\tau)$ is an ageing function.

For constants σ_i and σ, the following relations are obtained from formulae (2.26):

$$\left. \begin{array}{c} \varepsilon_i(t) = \dfrac{\sigma_i}{G_m} + \sigma_i \displaystyle\int_{\tau_1}^{t} K_\gamma(t, \tau) d\tau; \\[3em] \varepsilon_v(t) = \dfrac{\sigma}{K_m} + \sigma \displaystyle\int_{\tau_1}^{t} K_v(t, \tau) d\tau. \end{array} \right\} \qquad \dots (2.27)$$

Upon differentiating the above equations, it is found that

$$\dot\varepsilon_i = \sigma_i K_\gamma(t, \tau); \quad \dot\varepsilon_v = \sigma K_v(t, \tau).$$

Consequently, when the expressions for creep kernels $K_\gamma(t, \tau)$ are $K_v(t, \tau)$ are known, the creep strain of soil can be described in the three-dimensional state of stress. These functions can be determined from the results of tests under uniaxial and triaxial compression and pure shear.

The hereditary creep theory developed by N.Kh. Arutyunyan is also able to take into account the process of hardening (ageing) of clayey soil with time with the help of the ageing function $\varphi(\tau)$. In this case, the creep kernel can be represented in the form:

$$K_\gamma(t, \tau) = \frac{\partial}{\partial \tau} \varphi_\gamma(\tau) \{ 1 - \exp[-\delta_\gamma(t - \tau)] \};$$

$$K_v(t, \tau) = \frac{\partial}{\partial \tau} \varphi_v(\tau) \{ 1 - \exp[-\delta_v(t - \tau)] \},$$

where δ_γ and δ_v are experimentally determined parameters:

$$\varphi_\gamma(\tau) = \frac{1}{G_2}\{1 - \exp[-\delta_\gamma(t - \tau)]\} + \frac{1}{G_3(\tau/\tau_1)};$$

$$\varphi_v(\tau) = \frac{1}{K_2}\{1 - \exp[-\delta_v(t - \tau)]\} + \frac{1}{K_3(\tau/\tau_1)};$$

Here $1/[G_3(\tau/\tau_1)] = a_\gamma \ln(\tau/\tau_1)$; $1/[K_3(\tau/\tau_1)] = a_v \ln(\tau/\tau_1)$.

The following equations can be employed to obtain the stress components:

$$\left.\begin{array}{l} \dot{\varepsilon}_x = (\sigma_x - \sigma)K_\gamma(t, \tau) + \sigma K_v(t, \tau); \ldots; \\ \dot{\varepsilon}_{xz} = \tau_{xz}K_\gamma(t, \tau). \end{array}\right\} \quad \ldots (2.28)$$

It may be mentioned that at present a number of modifications and versions are available for describing the creep kernel under shear and volumetric strain.

In engineering practice, the equation of hereditary creep is often used to describe the process of unidimensional consolidation (compression):

$$\left.\begin{array}{l} \varepsilon_z(t) = \sigma_z(t)m_{v1} - \int_{\tau_1}^t \sigma_z(\tau)\frac{\partial}{\partial\tau}m_v(t, \tau)d\tau, \\ m_v(t, \tau) = m_{v1} + m_{v2}[1 - e^{-\delta_{com}(t-\tau)}] + m_{v3}(\tau/\tau_1), \end{array}\right\} \quad \ldots (2.29)$$

where m_{v1} is the coefficient of stipulated instantaneous relative compression; m_{v2} is the coefficient of secondary relative compression, $m_{v3} = a_{com} \ln(\tau/\tau_1)$; δ_{com} is creep parameter under compression; a_{com} and τ_i are ageing parameters (aged creep).

Thus, in the simplest case of unidimensional consolidation, the hereditary creep equation contains five parameters: m_{v1}, m_{v2}, m_{v3}, τ_1 and δ_{com}, all of which can be readily determined from the results of compression tests on a single specimen under stepped loading.

The above equations of hereditary creep were examined for linear stress-strain relation. In the theory of hereditary creep, it is also possible to consider non linear stress-strain relation, by replacing the stresses and strains in eqn. (2.26) with the functions governing the nature of the stress-strain relation, i.e.,

$$\left.\begin{array}{l} \varepsilon_i(t) = \frac{1}{G_0}\{\sigma_i(t) + \int_0^t K_\gamma(t, \tau)f_1[\sigma_i, (\tau)]d\tau\}; \\ \varepsilon_v(t) = \frac{1}{K_0}\{\sigma(t) + \int_0^t K_v(t, \tau)f_2[\sigma(\tau)]d\tau\}. \end{array}\right\} \quad \ldots (2.30)$$

The theory of hereditary creep can be applied to soils only if a large number of rheological parameters are known. These parameters have to be determined experimentally. However, the larger the number of parameters, the higher the number of factors accounted for in describing the extremely complex process of creep in clayey soils and, correspondingly, the higher the reliability of the engineering predictions. The difficulties encountered in mathematical computations can be overcome by using numerical methods and computers.

Engineering theory of creep

The engineering theory of creep includes the theories of ageing, flow and hardening.

According to the theory of ageing, strain ε is equal to the sum of elastic and viscous strains, i.e., $\varepsilon = \varepsilon^e + \varepsilon^v$. A functional relation can be established between stress, creep strain and time for constant density and moisture content of soil: $f(\varepsilon^v, \sigma, t) = 0$, wherein

$$\varepsilon^v = \varphi(\sigma)\psi(t). \qquad \qquad \dots (2.31)$$

When there is a change in moisture content of soil (for example, in loess and swelling soils), its effect can be taken into account by introducing an additional argument on the right-hand side of formula (2.31). For similar creep curves, $\psi(0) = 1$. However, if the instantaneous strain does not satisfy the similitude conditions, then

$$\varepsilon = \sigma/E + \varphi(\sigma)\psi(t); \quad \psi(0) = 0, \qquad \qquad \dots (2.32)$$

where E is the modulus of elasticity.

In soil rheology [1, 3, 14, 17], functions of the type $\varphi(\sigma)$ have found maximum application:

$$\sigma = A\varepsilon^m; \quad \sigma = \frac{E_0\sigma^*\varepsilon}{\sigma^* + E_0^\varepsilon},$$

where A, $m \leq 1$ are parameters, E_0 is initial modulus of compression and σ^* is the ultimate strength under compression.

The following functions have been proposed by S.S. Vyalov for the time-dependent function $\psi(t)$.

$$\psi(t) = 1 + \delta \left(\frac{t}{T}\right)^\alpha; \quad \psi(t) = 1 + \delta \ln \frac{t+T}{T};$$

$$\psi(t) = 1 + (\delta - 1)\frac{t}{t+T},$$

where δ, $0^\dagger < \alpha \leq 1$ are parameters; t is time duration in sec and $T = 1$ sec.

The above theory has found wide application because of its simplicity. However, it is not invariant in respect of reference time of commencement and is only applicable, therefore, if the load is constant or varies monotonously.

---

† This appears to be a printing error in the Russian original—General Editor

In the flow theory, a functional relation is established between stress, rate of creep strain and time: $\varphi(\dot{\varepsilon}^v, \sigma, t) = 0$. For similar creep curves, the following relation is valid: $\dot{\varepsilon}^v = \varphi_1(\sigma)\psi_1(t)$. It is possible to describe damping steady or progressive yielding by selecting a suitable, $\psi_1(t)$ function.

Theory of plastic flow

This theory is based on the results of laboratory tests of the past few years and the modern theory of strain hardening a plastic medium [1, 2, 5, 17]. The latter is an extension of the theory of associated flow rule that generalises the flow theory by introducing the concept of plasticity potential ψ. The partial derivative of ψ is proportional to the increment of plastic strain:

$$d\varepsilon_{ij}^p = d\lambda \partial\psi/\partial\sigma_{ij}, \qquad \ldots (2.33)$$

where $d\lambda \geq 0$ is some infinitely small scalar multiplier.

In the stress space σ_{ij} of an ideal plastic medium there exists a yield surface $f(\sigma_{ij}) = K$ (where $K > 0$) that restricts the region of elastic strain described by $f < K$. Plastic flow occurs at stresses that lie on the yield surface.

If it is assumed that the yield function and plastic potential are the same ($\psi = f$), then $d\varepsilon_i^p = d\lambda \partial f/\partial\sigma_{ij}$. This indicates that plastic flow develops along the normal to the yield surface $f(\sigma_{ij}) = K$. If the Von Mises yield criterion $f(\sigma_{ij}) = T^2$ is applied, then the material will be in elastic state when $T < \tau$ and in plastic state when $T = \tau_s$. In the space of principal stresses the above equation describes the surface of a cylinder with axis $\sigma_1 = \sigma_2 = \sigma_3$.

Equation (2.33) is known as the associated plastic flow rule because it is related (associated) with yield condition. By means of this rule, the general plasticity equation can be examined by introducing yield surfaces of various shapes. However, for soil medium, although the shape of yield surface has been studied in detail, it is not as significant as the shape of load surface, because the latter allows the plastic strain to be estimated at stresses less than the yield limit.

This problem can be solved on the basis of the theory of strain hardening a plastic medium, which is founded on the concept of load surface Σ, which separates the region of elastic deformation from the region of plastic deformation in stress space σ_{ij}. An infinitely small stress increment $d\sigma_{ij}$ (additional loading) results in either elastic deformation or a continuous plastic deformation. Unloading is observed to occur when $d\sigma_{ij}$ is directed inward into Σ, resulting in elastic deformation; $d\sigma_{ij}$ directed outward from Σ represents loading resulting in plastic deformation. Finally, if $d\sigma_{ij}$ is directed along the tangent to the stress surface, it represents neutral loading which again results in elastic deformation. The last condition is a prerequisite for a continuous transition from plastic deformation to elastic and vice versa, when there is a continuous variation of the direction of load vector $d\sigma_{ij}$. The nature and magnitude of surface load depends on the current state of stress, the history of soil formation and, of course, the density of soil skeleton. It may be mentioned that for soil medium, the load surface is a closed one because plastic deformation is also produced by all-round hydrostatic compression.

At present, there are a number of theories of plastic flow with hardening based on various shapes and dimensions of load surfaces Σ. The components of stress tensor, plastic strains, density of soil and its moisture content serve as arguments of the load function. If the load function around the point under consideration can be differentiated with respect to σ_{ij}, then by virtue of this fact it has a unique normal and is, therefore, a regular function. Otherwise, it is a non-regular function. A load surface having features such as ribs and corners can be described by means of piecewise smooth functions.

It has been demonstrated by Yu.K. Zaretskii and V.N. Lombardo [5] that the most general description of the behaviour of soil media under arbitrary load is possible by applying the theory of plastic flow with hardening, based on V. Koiter's associated flow rule.

4. Plastic Flow of Strain Hardening Clayey Soils

Plastic deformation of clayey soils in the condition preceding the state is distinguished by its lag with respect to the time of application of load increment. This is equally true for shearing as well as volumetric deformation. However, damping at the rate of shearing plastic strain is invariably slower.

With the passage of a certain period $(t = t^*)$, during which strains stabilise, given stress increments $\Delta \tau$ and $\Delta \sigma$ will be accompanied by increments of plastic strain:

$$\Delta \gamma^p = \int_0^{t^*} \dot{\gamma}^p dt; \quad \Delta \varepsilon^p = \int_0^{t^*} \dot{\varepsilon} dt. \qquad \ldots (2.34)$$

It follows from the foregoing that in developing a theory of plastic flow with strain hardening for clayey soils, it is important to identify the mechanism by which the shearing and volumetric plastic strains grow and to describe them correctly, taking into account the governing factors.

The build-up of plastic strain in soil medium occurs chiefly by the shearing mechanism (mutual displacement of particles and aggregates), which is significantly complicated by volumetric plastic strain and thixotropic hardening. These features distinguish soil medium from other media and significantly influence both the methods employed for studying the growth of plastic strain as well as methods adopted for their description.

The results of tests on strength and deformability of clayey soils are conventionally treated as the source of initial values of density and moisture content. This approach is valid if the volumetric strain of soil and its moisture content vary over a small range (1 to 2%) in the course of deformation. However, volumetric strain may frequently be as high as 10 to 20%, which results in hardening of soil, thereby making it essential to adopt a special approach for describing the test results. As indicated earlier, if hardening is considered while examining the shearing process, the cohesion may vary manifold and the angle of internal friction by 8 to 10°. The theory of plastic flow of hardening clayey soil is based on the application of the slip plane theory to the planes of maximum shear deformation, whose orientations depend on the angle of internal friction. It is thus obvious that hardening due to consolidation and the thixotropic processes must be taken into consideration.

An analysis of the results of recent experimental studies revealed that maximum shearing and volumetric strains occur at a particular orientation of the load vector in stress space, i.e., when clayey soil is characterised by deformation anisotropy. Maximum shearing and volumetric strains occur when the load vector makes an angle $\pi/2 + \varphi$ and φ respectively with the σ axis (Fig. 2.9). When shearing strain is maximum, the volumetric strain is absent, and vice versa.

Let us examine the shearing and volumetric plastic strain of hardening clayey soil in the condition preceding the limit state (in terms of strength), which is caused by relative slip between soil particles and aggregates within the framework of the incremental plastic flow theory. The stabilised stress-strain state of soil will be examined first under stepped loading, assuming that after the application of load step $\Delta\sigma$ and $\Delta\tau$, the soil will experience elastic and viscoplastic strains that damp with time. The working hypothesis adopted for describing the mechanism of plastic strain growth is based on the scheme of relative sliding of individual aggregates over the slip plane, occupying a

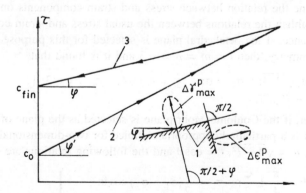

Fig. 2.9. Diagram depicting the growth of plastic strain in clayey soil.

1—orientation of the planes of maximum shearing and volumetric strain; 2—limit curve in case of loading; 3—limit curve in case of unloading.

38

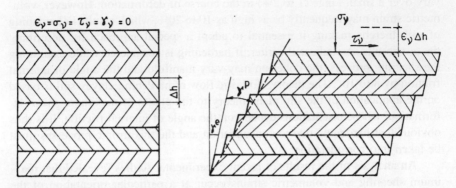

Fig. 2.10. Mechanism of growth of shearing plastic strain in an elementary volume of hardening clayey soil, based on the slip plane theory.

particular orientation in the stress space σ_{ij} (Fig. 2.10). The normal to this plane has direction cosines l, m and n. The shearing stresses τ_ν and normal stresses σ_ν on the plane are related to stress and strain components by the following expressions:

$$\left.\begin{array}{l} \sigma_\nu = \sigma_1 l^2 + \sigma_2 m^2 + \sigma_3 n^2; \\ \tau_\nu^2 = (\sigma_1 - \sigma_2)^2 l^2 m^2 + (\sigma_2 - \sigma_3)^2 m^2 n^2 + (\sigma_3 - \sigma_1)^2 l^2 n^2; \\ l^2 + m^2 + n^2 = 1; \sigma_1 \geq \sigma_2 \geq \sigma_3. \end{array}\right\} \quad \dots (2.35)$$

The relation between shearing and volumetric strains (ε_ν, γ_ν, ε_1, ε_2, ε_3, and also $\dot{\varepsilon}_\nu$, $\dot{\gamma}_\nu$, $\dot{\varepsilon}_1$, $\dot{\varepsilon}_2$, $\dot{\varepsilon}_3$) is given by a similar expression. For a particular orientation of the plane of maximum shear in the prelimit state, it is possible to first determine the relation between stress and strain components on this plane and then establish the relations between the usual stress and strain components.

For instance, if the octahedral plane is selected for this purpose in view of the axial symmetry, then $l = m = n = \sqrt{3}$ and it is found that

$$\left.\begin{array}{l} \tau_\nu = \sigma_i \sqrt{2/3}; \quad \sigma_\nu = \sigma; \quad \gamma_\nu = \varepsilon_i / \sqrt{6}; \quad \varepsilon_\nu = \varepsilon_\nu / 3; \\ \varepsilon_i / \sigma_i = 2\gamma_\nu / \tau_\nu. \end{array}\right\} \quad \dots (2.36)$$

However, if the Coulomb-Mohr plane is selected as the plane of maximum shear, which is a particularly appropriate choice for two-dimensional problems, then $m = 0$, $n = \sqrt{1 - l^2}$, $l = \cos\theta$ and the following relations are obtained:

$$\left.\begin{array}{l} \theta = \dfrac{\pi}{4} \pm \dfrac{\varphi}{2}; \quad \tau_\nu = \dfrac{1}{2}(\sigma_1 - \sigma_3)\cos\theta; \\[2mm] \sigma_\nu = \dfrac{\sigma_1 + \sigma_2}{2} + \dfrac{\sigma_1 - \sigma_2}{2}\sin\theta; \\[2mm] \gamma_\nu = \dfrac{\varepsilon_1 - \varepsilon_2}{2}\cos\theta; \varepsilon_\nu = \varepsilon_1 \cos^2\theta + \varepsilon_2 \sin^2\theta. \end{array}\right\} \quad \dots (2.37)$$

The transition from the general relations between stress and strain invariants to the relations between their components can be achieved by applying the condition of coaxiality of the vectors of stress and strain increments, i.e., the condition of similitude between the states of stress and strain expressed as $\mu_\sigma = \mu_\varepsilon$:

$$\frac{\Delta\varepsilon_x - \Delta\varepsilon_y}{\Delta\sigma_x - \Delta\sigma_y} = \cdots = \frac{\Delta\varepsilon_1 - \Delta\dot{\varepsilon}_2}{\Delta\sigma_1 - \Delta\sigma_2} = \cdots = \frac{\Delta\gamma_{xy}}{2\Delta\tau_{xy}} = \cdots = \frac{3\Delta\varepsilon_i}{2\Delta\tau_i}\chi_\gamma;$$

$$\dots (2.38)$$

where

$$\chi_\gamma = \frac{3\varepsilon_i}{2\tau_i} = \frac{f(\sigma_i, \sigma, t, K_\sigma)}{2\tau_i}; \quad K_\sigma = \frac{\Delta\sigma}{\Delta\tau}.$$

On the basis of formulae (2.38), the strain components can be determined for a given loading trajectory by Hencky's relations:

$$\Delta\varepsilon_x - \Delta\varepsilon = \chi_\gamma(\Delta\sigma_x - \Delta\sigma); \dots; \Delta\gamma_{xy} = 2\chi_\gamma\Delta\tau_{xy}. \qquad \dots (2.39)$$

On introducing the notation $\Delta\varepsilon = \chi_v\Delta\sigma$, eqns. (2.39) can be written in the form:

$$\left.\begin{aligned}
\Delta\varepsilon_x &= \chi_\gamma(\Delta\sigma_x - \Delta\sigma) + \chi_v\Delta\sigma; \\
\Delta\varepsilon_y &= \chi_\gamma(\Delta\sigma_y - \Delta\sigma) + \chi_v\Delta\sigma; \\
\Delta\varepsilon_z &= \chi_\gamma(\Delta\sigma_z - \Delta\sigma) + \chi_v\Delta\sigma; \\
\Delta\gamma_{xy} &= 2\chi_\gamma\Delta\tau_{xy}; \\
\Delta\gamma_{yz} &= 2\chi_\gamma\Delta\tau_{yz}; \\
\Delta\gamma_{zx} &= 2\chi_\gamma\Delta\tau_{zx}.
\end{aligned}\right\} \qquad \dots (2.40)$$

Consequently, unlike the existing plastic flow theories, in the theory described here it is possible to concomitantly take into account the dilatationary hardehing, loading trajectory parameter K_σ and orientation of the selected slip plane.

Relative slip between soil aggregates (see Fig. 2.10) on the plane under consideration is possible, provided the following condition is satisfied in a particular time interval.

$$\tau_v > \sigma_v \operatorname{tg}\varphi + c(t), \qquad \dots (2.41)$$

or

$$\tau_v + \Delta\tau_v > (\sigma_v + \Delta\sigma_v)\operatorname{tg}\varphi + c + \Delta c(t), \qquad \dots (2.42)$$

where φ is the angle of internal friction and $\Delta c(t)$ is cohesion that varies with time due to consolidation and structural changes.

As the plastic strain due to change of volume and shape builds up, the soil hardens and cohesion increases by Δc (Fig. 2.11), resulting in a stabilised state of strain:

$$\tau_v + \Delta\tau_v = (\sigma_v + \Delta\sigma_v)\operatorname{tg}\varphi + c + \Delta c. \qquad \dots (2.43)$$

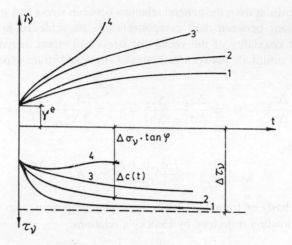

Fig. 2.11. Hardening of clayey soil under stepped loading.

1, 2, 3, 4 are curves corresponding to the first, second, third and fourth stages of loading, following a given load trajectory.

In the case of stepped loading this process may continue until the hardening capacity of soil is fully exhausted, leading to the state of limiting equilibrium.

Let us consider the shearing diagram for a given loading trajectory. In the elementary volume of soil under consideration, the relation between stress and strain may be represented in incremental form as it allows for elastic unloading and subsequent plastic deformation.

With the foregoing in view, the incremental moduli of distortion G^t, G^e, G^p and incremental bulk moduli K^t, K^e, K^p are now introduced. Superscripts t, e and p represent tangent, elastic and plastic moduli respectively. The first two distortion moduli are determined for the loading and unloading curves respectively with small loading and unloading steps. The third term is determined through the first two using the following well-known formulae:

$$1/G^t = 1/G^e + 1/G^p; \quad 1/K^t = 1/K^e + 1/K^p. \qquad \ldots (2.44)$$

It follows from the above that:

$$G^t = G^p/(1 + G^p/G^e), \text{ and } K^t = K^p/(1 + K^p/K^e).$$

Obviously, in the stabilised state, each loading step will produce an increment of elastic and plastic strains $\Delta\gamma = \Delta\gamma^e + \Delta\gamma^p$ and $\Delta\varepsilon = \Delta\varepsilon^e + \Delta\varepsilon^p$. The increments of plastic strain may be determined, provided the rates of change of the plastic strains $\dot{\gamma}^p$ and $\dot{\varepsilon}^p$ are known. We shall return to this aspect again.

Suppose that $\Delta\gamma^p$ and $\Delta\varepsilon^p$ are known and depend upon the state of stress, density and moisture content of soil, cohesion and angle of internal friction. It is then possible to write the following expressions:

$$\left.\begin{array}{l} \Delta\gamma_\nu(t^*) = \Delta\tau_\nu/G^t = \Delta\tau_\nu(1/G^e + 1/G^p); \\ \Delta\varepsilon_\nu(t^*) = \Delta\sigma_\nu/K^t = \Delta\sigma_\nu(1K^e + 1/K^p), \end{array}\right\} \qquad \ldots (2.45)$$

where

$$\left.\begin{array}{l} G^p = f(\sigma_\nu, \tau_\nu, K_\sigma, \mu_\sigma, t^*); \\ K^p = f(\sigma_\nu, \tau_\nu, K_\sigma, \mu_\sigma, t^*). \end{array}\right\} \qquad \ldots (2.46)$$

As a first approximation, it can be assumed that $G^e = \text{const.}$ and $K^e = \text{const.}$ are both unloading moduli and are independent of σ_ν, τ_ν and K_σ.

M.V. Proshin tested specimens of unsaturated clayey soil of given density and moisture content on a triaxial, unsymmetrical compression set-up and found that expressions (2.46) were complex in nature. These expressions can be represented in the form of curves, tables and empirical formulae:

$$\left.\begin{array}{l} G^l = \dfrac{G^e(\sigma_{i(1)} - \sigma_{i(2)})}{\dfrac{\sigma_{i(2)}}{(1 - \sigma_{i(2)}/\sigma_{i(2)}^*)^m} - \dfrac{\sigma_{i(1)}}{(1 - \sigma_{i(1)}/\sigma_{i(1)}^*)^m}}; \\[4mm] K^t = \dfrac{K^e(\sigma_{(2)} - \sigma_{(1)})}{\dfrac{\sigma_{(2)}}{1 + n\sigma_{(2)}} - \dfrac{\sigma_{(1)}}{1 + n\sigma_{(1)}}}. \end{array}\right\} \qquad \ldots (2.47)$$

We can now return to the initial equations (2.34) and determine the rate of growth of the shearing and volumetric viscoplastic strains. For a given load step, the rate of shearing plastic deformation is governed by the relation

$$\eta^{\nu p}\dot{\gamma}_\nu^P = \Delta\tau_\nu - \Delta\sigma_\nu \, \text{tg} \, \varphi - \Delta c(t), \qquad \ldots (2.48)$$

where

$$\dot{\gamma}_\nu^P = 0 \text{ or } \Delta\tau_\nu < \Delta\sigma_\nu \, \text{tg} \, \varphi + \Delta c(t) \text{ and } \dot{\gamma}_\nu^P > 0 \text{ or } \Delta\tau_\nu > \Delta\sigma_\nu \, \text{tg} \, \varphi + \Delta c(t).$$

The hardening of soil due to increase in cohesiveness may be taken into account on the one hand, through shearing and volumetric strains and, on the other, by thixotropic phenomena, i.e.,

$$\Delta c(t) = K_\varepsilon \Delta\varepsilon_\nu^P(t) + K_\gamma \Delta\gamma^P(t) + K(t)(1 - e^{-\lambda t}), \qquad \ldots (2.49)$$

where K_ε is the coefficient of dilatationary hardening which is determined experimentally; $K_\varepsilon = \Delta c/\Delta\varepsilon_\nu^P$; K_γ is the coefficient of hardening due to shearing which depends on the accumulated angular displacement; K_t is the coefficient of thixotropic hardening which is determined experimentally; $\lambda = \lambda_1 - \lambda_2$ (here λ_1 and λ_2 are the rates of restoration and breakdown of contacts in the soil).

If $\lambda_1 > \lambda_2$, the soil experiences hardening and the creep process becomes damped; if $\lambda_1 < \lambda_2$, the soil strength weakens and the creep process progresses further.

The variation of volumetric strain with time is described by the following relations of damping behaviour:

$$\left.\begin{array}{l} \Delta\varepsilon_v^P(t) = \Delta\varepsilon_v^P(\infty)(1 - e^{-\delta_v t}); \\ \Delta\varepsilon_v^P(\infty) = \Delta\sigma/K^P(\sigma), \end{array}\right\} \qquad \ldots (2.50)$$

where δ_v is the volumetric creep parameter.

The dilatationary strain is added to the volumetric strain given above, i.e.,

$$\Delta\varepsilon_v^P(t) = \Delta\varepsilon_v^P(t)(1 - e^{-\delta t}) + d\gamma^P(t), \qquad \ldots (2.51)$$

where d is dilatation parameter.

Assuming $K_\gamma = $ const. and substituting (2.51) first in formula (2.49) and then in (2.48), it is found that:

$$\dot{\gamma}_v^P + f_1\gamma_v^P = f_2(t), \qquad \ldots (2.52)$$

where

$$f_1 = \frac{K_\varepsilon d + K_\gamma}{\eta^{vp}} = a;$$

$$f_2(t) = \frac{\Delta\tau_v - \Delta\sigma_v \operatorname{tg}\varphi - \Delta\sigma_v(K_\varepsilon/K^P)}{\eta^{vp}}(1 - e^{-\delta t}) - \frac{K_t(1 - e^{-\lambda t})}{\eta^{vp}}.$$

The general integral of the first order, linear differential equation given above is:

$$\gamma_v^P = e^{-at}\left(\int f_2(t)e^{at}\,dt + C\right). \qquad \ldots (2.53)$$

If the indefinite integral in the above formula is replaced by a definite integral in the limits t_0 to t, then the value of C can be determined at $t = t_0$.

For the simplest case, when $\eta^{vp} = $ const., it is found that

$$\gamma_v(t) = \frac{\Delta\tau_v - \Delta\sigma_v \operatorname{tg}\varphi}{\eta^{vp}a}(1 - e^{-at}) + \frac{\Delta\sigma_v}{\eta^{vp}}\frac{K_\varepsilon}{K_p} \times \left(\frac{1 - e^{-at}}{a} - \frac{e^{-\delta t} - e^{-at}}{a - \delta}\right)$$

$$- \frac{K_t}{\eta^{vp}}\left(\frac{1 - e^{-at}}{a} - \frac{e^{-\lambda t} - e^{-at}}{a - \lambda}\right). \qquad \ldots (2.54)$$

It is evident that, depending on the sign of λ in the above formula, the creep process is either damped or progresses further.

In case of damping creep ($\lambda > 0$), at $t \to \infty$, we have,

$$\gamma_v^P(\infty) = \frac{1}{K_\varepsilon d + K_\gamma}[\Delta\tau_v - \Delta\sigma_v(\operatorname{tg}\varphi + K_\varepsilon/K^P) - K_t]. \qquad \ldots (2.55)$$

It follows from the above, that the incremental modulus of shearing plastic strain is given by the expression

$$G_p = \frac{(K_\varepsilon d + K_\gamma)\Delta\tau_v}{\Delta\tau_v - \Delta\sigma_v(\operatorname{tg}\varphi + K_\varepsilon/K^P) - K_t}. \qquad \ldots (2.56)$$

It may be thus concluded that the modulus of shearing plastic strain depends on the stress increments $\Delta\sigma_\nu$ and $\Delta\tau_\nu$ (i.e., on loading trajectory), hardening parameters K_ε, K_γ and K_t, dilatation parameter d, angle of internal friction φ and incremental modulus of volumetric plastic strain $K^p(\sigma_\nu)$. Further, when $\Delta\tau_\nu \leq \Delta\sigma_\nu(\text{tg }\varphi + K_\varepsilon/K^p) + K_t$, then $G^p \to \infty$ and the soil does not experience deformation. On the other hand, when $\Delta\tau_\nu > \Delta\sigma_\nu(\text{tg }\varphi + K_\varepsilon/K^p) + K_t$, then $\infty > G^p > 0$ and the soil experiences a certain increment of shearing plastic strain.

As K_ε, K^p and K_γ depend upon the initial stressed state represented by τ and σ as also the extent to which it approaches the limiting state, expression (2.56) should obviously be examined within the limits of the given load step. For example, as $\gamma \to \gamma^*$, the effect of hardening on the creep process continuously weakens; therefore, in eqn. (2.56) it is possible to assume $K_\varepsilon = K_\gamma = 0$, wherefrom it ensues that $G^p = 0$, i.e., the soil experiences transition into the state of flow.

If it is assumed that in the prelimit state the planes of maximum shear have the same orientation as the slip planes in the limit state, then it is possible to interpret the theory described above in a number of ways. For instance, if the octahedral plane is selected as the plane of maximum shear, then $l = m = n = 1 = 1/\sqrt{3}$ and the following relations hold good:

$$\left.\begin{array}{ll} \tau_\nu = \sqrt{2}\sigma_i/\sqrt{3}; & \sigma_\nu = \sigma; \\[2mm] \gamma_\nu = \varepsilon_i\sqrt{6}; & \varepsilon_\nu = \varepsilon_v/3. \end{array}\right\} \qquad \dots (2.57)$$

If the plane determined from the Coulomb-Mohr strength theory is selected as the plane of maximum shear, then it is found that $m = 0$, $n = \sqrt{1 - e^2}$, $e = \cos\theta$; $\theta = \pi/4 \pm \varphi/2$. In this case,

$$\left.\begin{array}{l} \tau_\nu = \dfrac{1}{2}(\sigma_1 - \sigma_3)\cos\varphi; \\[3mm] \sigma_\nu = \dfrac{\sigma_1 + \sigma_3}{2} + \dfrac{\sigma_1 - \sigma_3}{2}\sin\varphi; \\[3mm] \gamma_\nu = \dfrac{\varepsilon_1 - \varepsilon_3}{2}\cos\varphi; \quad \varepsilon_\nu = \varepsilon_1\cos^2\theta + \varepsilon_3\sin^2\theta. \end{array}\right\} \qquad \dots (2.58)$$

In order to graduate from the general relations between stress and strain invariants to relations between their components, it is necessary to apply the condition of co-axiality between the vectors of stress and strain increments expressed as $\mu_\varepsilon = \mu_\sigma$. Consequently, it is possible to use Hencky's relations (2.39). By separating elastic strain from permanent strain, it is possible to introduce the notation $\chi = 1/(2G) + \lambda$, i.e.

$$\left.\begin{array}{l} \varepsilon_x = \dfrac{\sigma_x}{E} - \mu(\sigma_y + \sigma_z) + \lambda(\sigma_x - \sigma); \\[3mm] \gamma_{zx} = \dfrac{\tau_{zx}}{G} + 2\lambda\tau_{zx}. \end{array}\right\} \qquad \dots (2.59)$$

The first terms in the above expressions represent the elastic strain components, while the second terms represent plastic strain components; therefore:

$$\varepsilon_i^p = \frac{2}{3}\lambda \tau_i; \quad \lambda = \frac{3}{2}\frac{\varepsilon_i^p}{\sigma_i}.$$

Hence

$$\varepsilon_x^p = \frac{3}{2}\frac{\varepsilon_i^p}{\sigma_i}(\sigma_x - \sigma); \quad \gamma_{zx}^p = \frac{\varepsilon_i^p}{\sigma_i}\tau_{zx}. \qquad \ldots (2.60)$$

The problem has thus been fully solved.

It may be mentioned that in the above theory of plastic flow of hardening clayey soils it has been demonstrated that, in principle, it is possible to take into account some of the important parameters of the hardening process while compiling the equations of state. An unwieldy equation was obtained despite the fact that only a limited number of parameters were considered and certain assumptions were made regarding their constancy. This again confirms how complex is the mechanism governing the creep process. In this theory almost all the real soil properties and their governing parameters have been considered, viz., φ, c, η^{vp}, δ, K^p, K_ε, K_γ, K_t and K_σ. Hence, from the stand point of practical applicability, it is desirable to simplify the equation of state and reduce the number of parameters by replacing them with experimental parameters that have no particular physical meaning.

5. Consolidation of Clayey Soils

The main factors affecting the duration of the consolidation process are deformability and permeability of soil, the size and shape of the soil mass being consolidated (length of the drainage path), the nature of applied forces etc.

If the soil skeleton has distinct rheological properties, the soil water contains dissolved and undissolved gases and the coefficient of permeability is less than 10^{-7} cm/sec, then the consolidation process will be complex. For the sake of comparison, it may be mentioned that the Terzaghi-Gersevanov theory of consolidation by seepage was based on the simplest assumptions: the soil skeleton was considered to be elastic and pore water to be non-compressible.

For the sake of convenience, analysis of the consolidation process will be divided into three stages, which are distinguished by different stress-strain states.

The *first stage* covers the initial stress-strain state of soil mass when stress and strain fields have been produced in the soil skeleton and pore water after application of the external load, but the drainage process has not yet extended into the depth of the soil mass and hence may be neglected. This stage does not continue for long but does determine the initial stress-strain state of the soil mass and, consequently, governs the consolidation process.

The *second stage* pertains to the intermediate stress-strain state of soil mass during which the pore water is intensively expelled from the whole volume and the excess pore pressure is dissipated. This stage ends with the almost total dissipation of excess pore pressure, although the rate of strain of the skeleton is not yet zero. However, as the strain rates are small, reactive pore pressure is not produced, i.e., mutual permeation between the skeleton and pore water occurs without viscous resistance of water.

The *third stage* deals with the final stress-strain state of soil mass when the whole external load is resisted by the soil skeleton, which slowly deforms according to its rheological properties.

The initial stress-strain state is fully described by the total stresses and the reduced secant or tangent modulus of deformation of soil:

$$\left. \begin{aligned} K_{\text{red}} &= K_{sk} + K_{g,w}/n; \quad G_{\text{red}} = G_{sk}; \\ \mu_{\text{red}} &= \frac{K_{\text{red}} - 2G_{sk}}{2(K_{\text{red}} + G_{sk})}, \end{aligned} \right\} \qquad \ldots (2.61)$$

where K_{sk}, G_{sk} are the secant or tangent bulk modulus and the modulus of shear of skeleton respectively, $K_{g,w}$ is the coefficient of volumetric compressibility of water containing dissolved gases: $K_{g,w} = 3m_{g,w}$;

$$m_{g,w} = (1 - S_r'/S_r'')/\Delta u_w; \qquad \ldots (2.62)$$

here $m_{g,w}$ is the coefficient of relative compressibility of pore water with dissolved gases; S_r', S_r'' is the degree of saturation of soil before and after the change in pore pressure by Δu_w.

The reduced moduli of soil depend upon the compressibility of pore water. For example, when $K_{g,w} \to \infty$, i.e., when the pores are filled with degassified water ($K_{\text{red}} \to \infty$), $\mu_{\text{red}} \to 0.5$ and, on the whole, the soil is incompressible.

The initial distribution of pore pressure in the soil mass may be determined from the sum of total principal stresses by means of the condition $\varepsilon = n\varepsilon_w$ and eqn. (2.19):

$$u_w(x, y, z) = \frac{\sigma_v(x, y, z)}{3} \frac{K_{g,w}}{nK_{sk} + K_{g,w}} \qquad \ldots (2.63)$$

Hence, if the solution to the problem of distribution of total stresses in the soil mass is known, the initial distribution of pore pressure may be determined from formula (2.63).

It may be mentioned that the initial stress-strain state of soil mass represents the state when the process of distribution and re-distribution of total stresses between the skeleton and pore water is complete. The duration of this process depends mainly on volumetric creep of the skeleton and to a much less extent on the duration of the consolidation process, i.e., the duration of the second stage. In view of the above, the second factor may be neglected.

The intermediate stress-strain state is distinguished by intensive expulsion of water from the pores of the whole volume of the consolidating soil mass. It can be described by just the equation of consolidation:

$$\frac{\partial \varepsilon_v}{\partial t} - nm_{g,w}\frac{\partial u_w}{\partial t} = \frac{k}{\gamma_w}\nabla^2 u_w. \qquad \dots (2.64)$$

The left-hand side of eqn. (2.64) represents the change in volume of soil pores in unit time due to consolidation of the skeleton and compression of pore water. The right-hand side represents the flow rate of water in an elementary volume during the same period.

It is evident that different results will be obtained, depending on the equation employed for determining the relation between the rate of volumetric strain and the skeleton stresses varying with time.

Thus the following equation of consolidation is obtained if the simplest assumptions made in the Terzaghi-Gersevanov theory are applied, i.e., the skeleton is assumed to be elastic and the water to be incompressible:

$$\frac{\partial u_w}{\partial t} = c_v \frac{\partial^2 u_w}{\partial z^2}, \qquad \dots (2.65)$$

where $c_v = k/(\gamma_w m_v)$ (here m_v is the coefficient of relative compressibility of the skeleton).

If the skeleton is assumed to have linear hereditary creep [see formulae (2.27)] and pore water is assumed to be linearly compressible, then the following equation of consolidation is obtained:

$$\frac{\partial^2 u_w}{\partial t^2} + A\frac{\partial u_w}{\partial t} = c_v\left(\delta\frac{\partial^2 u_w}{\partial z^2} + \frac{\partial^2 u_w}{\partial z^2 \partial t}\right), \qquad \dots (2.66)$$

where

$$A = \frac{\delta(m_{v0} + nm_{g,w})}{m_{v0} + nm_{g,w}}; \quad c_v = \frac{k}{\gamma_v(m_{v0} + nm_{g,w})};$$

$$m_{v0} = m_{v1} + m_{v2}.$$

A comparison of eqns. (2.65) and (2.66) indicates how complex the consolidation process becomes if the real properties of clayey soil are taken into consideration, viz., skeleton creep and compressibility of pore water. The final solutions of eqns. (2.65) and (2.66) are given below for a layer of thickness h in the case of drainage at both ends:

$$u_w(z,t) = \frac{4q}{\pi}\sum_{n=1,3,\dots}^{\infty}\frac{l}{n}\sin\frac{\sigma nz}{h}e^{\lambda_n t}, \qquad \dots (2.67)$$

where $\lambda_n = -\pi^2 c_v n^2 / h^2$;

$$u_w(z,t) = \frac{4q}{\pi}\beta_0 \sum_{n=1,3,\ldots}^{\infty} \frac{1}{n}\left(\frac{B_0 - \lambda_2}{\lambda_1 - \lambda_2}e^{\lambda_1 t} + \frac{B_0 - \lambda_1}{\lambda_2 - \lambda_1} \times e^{\lambda_2 t}\right)\sin\frac{\pi n z}{h},$$

$$\ldots (2.68)$$

where

$$\beta_0 = \frac{m_{v1}}{m_{v1} + n m_w}; \quad B_0 = \frac{\delta m_{v2}}{m_{v1}}(1 - \beta_0);$$

$$\lambda_{1,2} = \frac{1}{2}\left[-\left(A + c_v\frac{\pi^2 n^2}{h^2}\right) \pm \sqrt{\left(A + c_v\frac{\pi^2 n^2}{h^2}\right)^2 + 4\delta c_v\frac{\pi^2 n^2}{h^2}}\right].$$

Equations (2.67) and (2.68) determine how the pore pressure decays in soil layer of thickness h under double-ended drainage. The two equations differ qualitatively (Fig. 2.12) although both describe the variation of stresses in the soil skeleton, i.e., $\sigma'(z,t) = q - u_w(z,t)$.

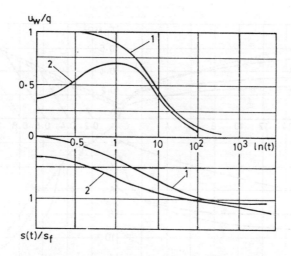

Fig. 2.12. Variation of pore pressure and settlement with time in a layer of clayey soil according to (1) the Terzaghi-Gersevanov theory of consolidation drainage and (2) the theory proposed by the author.

The equation given below is integrated by taking into consideration formula (2.68) for determining the settlement of soil layer, taking into account skeleton creep and compressibility of water:

$$s(t) = \int\limits_0^h \varepsilon_1(z,t)dz = \int\limits_0^h \left[m_{v1}\sigma_1(z,t) - \int\limits_0^h \sigma_1(z,\tau) \times \frac{\partial}{\partial t} m_{v0}(t,\tau)d\tau \right] dz$$

The following equation is obtained after integration:

$$s(t) = qh[m_{v1}u_{\mathrm{I}}(t) + m_{v2}u_{\mathrm{II}}(t)], \qquad \qquad \ldots (2.69)$$

where $u_{\mathrm{I}}(t)$ and $u_{\mathrm{II}}(t)$ are functions of m_{v1}, m_{v2}, k, $m_{g,w}$ and h^2.

The solution of consolidation equation (2.64) is further complicated if ageing of the soil skeleton is taken into account, resulting in hypergeometric functions [17, 21]. In view of the above, it has been proposed that the following additional term should be considered while predicting creep deformation due to ageing of the skeleton:

$$s(t) = qh[m_{v1}u_{\mathrm{I}}(t) + m_{v2}u_{\mathrm{II}}(t) + m_{v3}u_{\mathrm{III}}(t)], \qquad \qquad \ldots (2.70)$$

where m_{v3} is an empirical parameter and $u_{\mathrm{III}}(t) = \ln(t/t_f)$.

It is thus possible to describe in a relatively simple manner the third stage of consolidation of clayey soil or the so-called secondary consolidation. Pore pressure is absent in this stage and the soil settlement grows proportionately with the logarithm of time (Fig. 2.12).

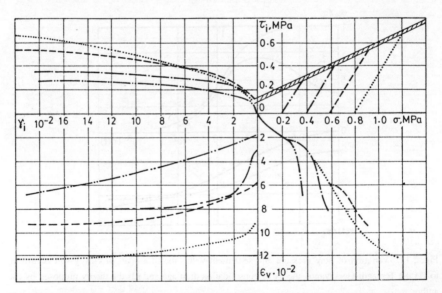

Fig. 2.13. Graphic record of stabilometric tests of red-brown clay with disturbed structure ($\gamma_s = 27.2$ kN/m^3, $\gamma = 19.6$ kN/m^3, $w = 0.23$).

In conclusion, the results of tests conducted on the triaxial compression set-up are presented below. It is evident from Figs. 2.13 to 2.15 that with time the volumetric and distortional strains cross the developed at the different functions. These properties are related for different types of clay and during their deformation will be discussed in terms of the general view. The author includes a large number of experimental results of the experimental problems in the solution of

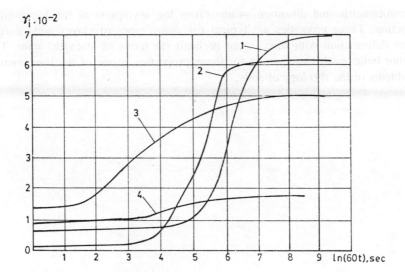

Fig. 2.14. Variation of shearing strain γ_i with time in semi-logarithmic scale (red-brown clay with disturbed structure wherein $\gamma_s = 27.2$ kN/m^3, $\gamma = 19.6$ kN/m^3, $w = 0.23$).

1—$\sigma = 0.4$ MPa and $\tau_i = 0.31$ MPa; 2—$\sigma = 0.34$ MPa and $\tau_i = 0.24$ MPa; 3—$\sigma = 0.78$ MPa and $\tau_i = 0.31$ MPa; 4—$\sigma = 0.22$ MPa and $\tau_i = 0.035$ MPa.

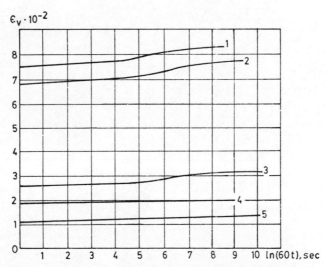

Fig. 2.15. Variation of volumetric strain ε_v with time in semi-logarithmic scale (red-brown clay with disturbed structure wherein $\gamma_s = 27.2$ kN/m^3, $\gamma = 19.6$ kN/m^3, $w = 0.23$).

1—$\sigma = 0.6$ MPa and $\tau_i = 0.346$ MPa; 2—$\sigma = 0.78$ MPa and $\tau_i = 0.31$ MPa; 3—$\sigma = 0.28$ MPa and $\tau_i = 0.14$ MPa; 4—$\sigma = 0.22$ MPa and $\tau_i = 0.035$ MPa; 5—$\sigma = 0.11$ MPa and $\tau_i = 0.017$ MPa.

the volumetric and distortion strains cross the asymptote of the logarithmic function. These properties are typical for many types of clayey soils during their deformation with time in the prelimit (in terms of strength) state. The author believes that accounting for these properties is one of the fundamental problems of the rheology of soils.

3

Theoretical Basis of Rheological Testing of Clayey Soils

1. Basic Principles of Rheological Testing of Clayey Soils

The rheological properties of clayey soils, like those of any other medium, represent the ability to offer viscous or viscoplastic resistance to changes in volume and shape. Such a definition of rheological properties of clayey soils is similar to that employed for describing the rheological properties of structural materials (concrete, timber, plastic, etc.). It does not reflect the specific features of a multiphase dispersive soil medium, consisting of a mineral skeleton, pore water and dissolved and undissolved gases.

A single-phase medium acquires its properties by molecular interaction, while the properties of the soil medium take shape due to interaction between the particles and phases. Therefore, the behaviour of single-phase media can be used to describe the rheological properties of clayey soils only in respect of the soil skeleton, as the density of the latter does not change much in the course of deformation.

Henceforth, the rheological behaviour of clayey soil will be understood to represent the ability of the soil skeleton to change its volume and shape under constant stresses acting in it. However, if the clayey soil is water saturated, its volumetric strain is governed, on the one hand, by creep of the skeleton and, on the other, by the rate at which water is expelled from the pores. In principle, the second process is also a rheological one, as it involves expulsion of a viscous fluid from the capillaries. This is the very factor that complicates the study and description of rheological properties of the skeleton of saturated soil under volumetric changes.

Unfortunately, one observes that even today the results of field and laboratory experiments on clayey soils are processed and analysed from the standpoint of the rheology of a single-phase medium, with no consideration of the degree of saturation. There is a risk in such cases of obtaining erroneous results due to the superposition of the consolidation process on the temporal process of deformation.

As the results of the tests are influenced by the experimental conditions and the initial physical properties (density and moisture content) of the clayey soil, it is desirable to select the test conditions and methods by taking into account the initial density and moisture content of the soil. For this purpose clayey soils are subclassified as quasi-single phase and quasi-double phase.

Quasi-single phase soils are those with an initial degree of water saturation $S_r < 0.8$, which varies during the deformation process but remains less than 0.8, thereby eliminating the possibility of development of residual pore pressure and consequent development of the consolidation process.

Quasi-double phase soils are those with an initial degree of water saturation $S_r \geq 0.8$, which varies in the course of deformation and may approach a value equal to unity, i.e., the soil may attain the state of complete water saturation. In this case, the development of excess pore pressure is unavoidable and, therefore, development of the consolidation process is inevitable.

It is also necessary to appreciate the role of test conditions. For example, the results of tests conducted on water-saturated clayey soils are significantly affected by the presence or not of free drainage. On the other hand, drainage conditions are of no consequence in testing unsaturated clays. Therefore, for quasi-double phase soil with and without preliminary consolidation, it is necessary to differentiate between drained and undrained test conditions.

As the tests conducted on quasi-double phase soils with drainage can produce a non-homogeneous state of stress, it is difficult to interpret the results of these tests in an unstabilised state, particularly for triaxial loading. It is advisable to test quasi-double phase soils after preliminary consolidation, with subsequent deviatory loading in the absence of drainage, accompanied by recording of pore pressure, i.e., carry out consolidated undrained tests. Preliminary consolidation of soil ensures that the state of stress of the specimen conforms with the initial state of stress of the soil mass. In addition, preliminary consolidation under maximum load is helpful in producing the same soil density for all tests under shear, despite the difference in the normal stresses.

2. Rheological Parameters of Clayey Soils

From the review of modern creep theories for soils presented in Chapter 1, it is clear that each of them deals with a large number of rheological parameters. In each particular case, the rheological parameters were deployed and determined in a different manner. A relation does not always exist between the rheological parameters used in the various theories. One can only guess about the similarity between parameters by the role they play in describing the creep process. However, some rheological parameters are common to all theories: elastic characteristics of soil G_m and K_m, determined from a stipulated value of instantaneous strain or magnitude of relieved load; viscosity η and η_v under shearing and volumetric deformation respectively; strength parameters φ, c_m, c_l and c_l and long-term strain parameters G_l and K_l.

Here attention will be devoted mainly to the methods of determining the above parameters and the rheological parameters of soil in conformity with the theories of hereditary creep and viscoplastic flow.

Parameters of the theory of hereditary creep of clayey soils

These parameters include the moduli of instantaneous strain G_m and K_m, moduli of long-term strain G_l and K_l, creep parameters δ_γ and δ_v for distortion and volumetric deformation respectively, ageing parameters τ_1, α_γ, α_v, and, finally, the parameters of the non-linear creep theory. At first glance, the number of parameters appears to be unnecessarily large. However, they describe only a single creep curve (relaxation) and may be determined by testing two identical specimens under shear and isotropic compression. For this, it suffices to know the initial stipulated instantaneous strain and the manner in which this strain develops with time.

In engineering practice, it is often found necessary to determine the rheological parameters of soil under constrained compression. In such cases, it suffices to consider only the following parameters: m_{v1}, m_{v2}, m_{v3}, δ_{com} and τ_1. Here m_{v1}, m_{v2}, m_{v3} are the coefficients of relative compressibility at time $t = 0$, $t = t_1$ and $t > t_1$ after application of the given load step, δ_{com} is the creep parameter under constrained compression and τ_1 is the parameter that describes ageing (hardening) of soil. A definite relation exists (see Chapter 2) between the creep parameters under constrained compression and compound stress-strain state.

Parameters of viscous flow of clayey soils

In the simplest case, soil may be considered a viscous compressible fluid and only the viscous flow parameters η and η_v may be employed for describing the rheological process. Such a state may be observed under small stresses acting over a long period during which the process proceeds towards the stage of slow, indefinite creep. This is important for structures that are very sensitive to deformation of the foundations and also rigid structures resting on slowly moving slopes.

The viscosity parameters η and η_v under distortion and volumetric strain respectively are related with viscosity under linear compression (expansion) by the following expression [1]:

$$\lambda = \frac{3\eta\eta_v}{\eta + \eta_v}. \qquad \qquad \dots (3.1)$$

If the relation for viscosity λ is expressed through the coefficient of lateral strain $\nu_v = \dot{\varepsilon}_2 / \dot{\varepsilon}_1$, then it is found that;

$$\eta_v = \frac{\lambda}{1 - 2\nu} = 2\eta \frac{1 + \nu_v}{1 - 2\nu_v} \eta. \qquad \cdots (3.2)$$

Let us compare the above expression with the following well-known relations:

$$K = 2G \frac{1 + \nu}{1 - 2\nu} = \frac{E}{1 - 2\nu}.$$

It can be seen from the comparison that the two sets of relations are identical and that the latter may be obtained from (3.2) by replacing K by η_v, G by η, E by λ and ν_v by ν.

The stress-strain state of such a medium can be described by the equations of state:

$$\dot{\varepsilon}_x - \dot{\varepsilon} = (\sigma_x - \sigma)/(2\eta), \ldots, \dot{\gamma}_{yz} = \tau_{yz}/\eta; \quad \dot{\varepsilon} = \sigma/\eta_v \qquad \cdots (3.3)$$

and the equilibrium equations which, in the given case, can be represented in the form given below:

$$\frac{\partial v_x}{\partial t} = F_x + \frac{1}{\rho} \frac{\partial \sigma}{\partial x} + \frac{\eta}{\rho} \left(\frac{\partial^2 v_x}{\partial x^2} + \frac{\partial^2 v_x}{\partial y^2} + \frac{\partial^2 v_x}{\partial z^2} \right) + \frac{1}{\rho}(\eta_v + \eta) \frac{\partial \varepsilon}{\partial x} \text{ etc.}$$

where ρ is density, v_x, v_y and v_z are components of the displacement velocities and σ is the mean stress.

These equations of motion of a viscous fluid are known as Navier-Stokes equations and constitute the fundamental equations of hydrodynamics of viscous fluids.

For a narrow range of variation of stresses, the parameters under consideration are not significantly affected by stresses but vary with time. In case of volumetric deformation, the viscosity should vary such that the volumetric strain decreases with time, i.e.,

$$\eta_v(t) = \eta_{v_0} e^{-\mu_v t}, \qquad \cdots (3.4)$$

where η_{v_0} is initial volumetric viscosity and μ_v is a parameter of the $\eta_v(t)$ curve.

Upon substituting the above relation in the expression $\dot{\varepsilon}_v = \sigma_v/\eta_v(t)$ and integrating, the following relation of decaying creep is obtained:

$$\varepsilon(t) = \frac{\sigma}{\eta_{v_0} \mu_v} (1 - e^{-\mu t}). \qquad \cdots (3.5)$$

The viscosity under shear also varies with time and its variation can be described by the following two equations, based on N.N. Maslov's theory and the ageing theory respectively.

$$\eta = \eta_\infty - (\eta_\infty - \eta_0)e^{-\mu t}, \qquad \cdots (3.6)$$

$$\eta = \eta_0(1 + \mu t), \qquad \cdots (3.7)$$

where η_∞ and η_0 are the final and initial value of viscosity respectively and $\mu(\text{sec}^{-1})$ is a decay parameter.

If the above expressions are substituted in the equation of viscous flow, i.e., $\dot{\gamma} = \tau/\eta$, then the following equations are obtained after integration:

$$\dot{\gamma}^{vp}(t) = \frac{\tau}{\eta_\infty} \left\{ t + \frac{1}{\mu} \ln \left[\frac{\eta_\infty - (\eta_\infty - \eta_0)e^{-\mu t}}{\eta_\infty} \right] \right\}; \qquad \dots (3.8)$$

$$\dot{\gamma}^{vp}(t) = \frac{\tau}{\eta_0 \mu} \ln(1 + \mu t). \qquad \dots (3.9)$$

If there is significant variation of skeleton density during the creep process, then this factor must be taken into consideration while determining the viscosity parameters.

Parameters of viscoplastic flow

These parameters include the angle of internal friction, φ, cohesion c_m, c_l and c_{res} and, finally, viscosities η^{vp} and η_v^{vp}. Viscoplastic flow is possible at stresses $\tau > \tau^*$, while viscous flow is possible at all values of τ.

The parameters c_m, c_l and c_{res} describe the variation of cohesiveness until the onset of failure, i.e., in principle, cohesion is a time-dependent quantity. Hence there is some doubt regarding the correct value of viscosity η^{vp} in view of the fact that it is determined from the following equation:

$$\eta^{vp} = [\tau - \tau^*(t)]/\dot{\gamma}.$$

It transpires that the variation of coefficient η^{vp} with time is associated not with the variation of soil viscosity, but with the variation of strength limit or threshold. However, this is a purely theoretical problem and will not be discussed further here.

Parameters of consolidation of clayey soil

In laboratory conditions, it is convenient to determine the consolidation parameters on a constrained compression set-up under one- or two-way drainage of the medium. If suitable instrumentation is available, the pore pressure is measured either at the base of the specimen or at its middle. It is advisable to use the hereditary creep equation (2.29) for uniaxial compression with total restraint on lateral expansion. Then it only remains to determine relative compressibility coefficients m_{v1}, m_{v2}, m_{v3}, creep parameters δ_{com}, coefficient of permeability k or coefficient of consolidation $c_v = k/[\gamma_w(m_{v1} + nm_w)]$ where n is soil porosity and m_w is coefficient of compressibility of pore water with dissolved gases, which is determined by the expression:

$$m_w = (1 - S_r)/(p_a + u_{w_0}), \qquad \dots (3.10)$$

where p_a is atmospheric pressure and u_{w0} is initial pressure in pore water.

In order to establish the role of skeleton creep in the progress of conso-lidation by seepage, it is necessary to determine consolidation index n_c by the formula:

$$n_c = \frac{\ln t_2 - \ln t_1}{\ln h_2 - \ln h_1}, \qquad \ldots (3.11)$$

where t_2 and t_1 are the time required for completion of consolidation by seepage in specimens of height h_2 and h_1 respectively under identical drainage conditions (on one or both sides).

If it is found that $n_c \approx 2$, then it can be surmised that skeleton creep is weak and its effect on the consolidation process is not significant. However, when $1 < n_c < 2$, the creep of soil skeleton is distinctly in evidence and its effect on the process of compression during consolidation by seepage is significant.

For highly compressible, weak and saturated clayey soils, it becomes necessary to determine the non-linear deformability and imperviousness of soil because these parameters exert a significant influence on the process of compression with time.

3. Methods of Rheological Testing of Clayey Soils

Rheological tests for clayey soils differ from the usual standard tests in that special attention is paid to the study of time-dependent processes. These processes are manifest to varying degrees depending on the test conditions and moisture content of the clayey soil. Moreover, the time-dependent processes proceed in different ways in the laboratory specimens and under field conditions.

Let us first examine the laboratory methods of testing quasi-single phase soils. In this case, it is possible to apply the existing methods of soil testing under static and kinematic loading. Obviously, relaxation tests should also be treated as tests conducted under kinematic loading.

Among the modern methods of testing the rheological properties of clayey soils in laboratory conditions, the following have found the widest application: triaxial axisymmetrical compression of cylindrical specimens (Fig. 3.1a); triaxial axisymmetrical compression or tension of dumb-bell shaped specimens (Fig. 3.1b); triaxial axisymmetrical compression/tension and torsion of thin-walled cylindrical specimens (Fig. 3.1c); triaxial axisymmetrical constrained compression and torsion of thin-walled cylindrical specimens (Fig. 3.1d); triaxial unsymmetrical compression of cubical specimens (Fig. 3.1e); direct shearing of cylindrical specimens (Fig. 3.1f); forced direct shearing of cylindrical specimens (Fig. 3.1g); shearing along external cylindrical surface of specimens by twisting of the upper half of the cylinder (Fig. 3.1h); ring shearing along a transverse plane of thin-walled cylindrical specimens (Fig. 3.1i) and multiplanar shearing of cylindrical specimens (distortion) (Fig. 3.1j).

In the modern methods of soil testing, it is possible to transmit forces, kinematic effects and a combination of the two to the specimen. However, it is

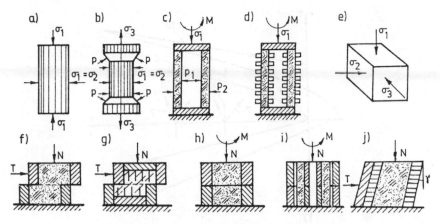

Fig. 3.1. Modern schemes of rheological testing of soils.

not always possible to specify the load condition $\sigma(t)$ and deformation condition $\Sigma(t)$ concomitantly in the various directions. For example, when the test is conducted in the consolidation set-up, a lateral pressure is produced on the stationary edge and it is not possible to control the loading trajectory. Fig. 3.2 depicts schematically how the various stress and strain trajectories are achieved in modern set-ups. Curve 1 represents the trace of the limit surface on the $\tau - \sigma$ plane; curve 2, the limiting surface of deformation; curve 3, the trajectory of uniaxial compression; curves 4 and 5, the trajectories of constrained compression; curves 6 and 8, the trajectories of simple shear; curve 7, the inclined trajectory of deviatory loading; curve 9, the trajectory of pure shear and curve 10, the limiting shear strength during unloading.

Static force may be applied in steps or in one stage depending on the objective of the tests. After application of a load step, strain development is monitored until such may be considered stable. A particular rate of growth of creep strain is adopted as the stabilisation criterion. These rates differ for volumetric and shearing strains.

A kinematic input may be applied on the soil specimen by means of a device that forces the specimen to deform at a given rate. The forces arising as a result of the above input are recorded by a dynamometer. The stress-strain rates may vary in the range 1 to 10^{-6} mm/min. This test makes it possible to establish the relation between the rate of creep strain and stresses in the prelimit state as well as in the limit state. It also helps in determining the residual strength after the peak strength has been attained. The scheme of slow deformation is the most difficult to implement. For conducting these tests one may either use low speed motors with large reduction gears or loading devices with electromagnetic step motors.

58

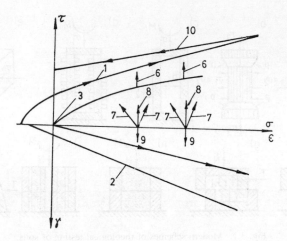

Fig. 3.2. Schematic representation of the various loading trajectories of soil specimen under compound state of stress.

Relaxation tests considerably reduce the time required to generate the necessary data for determining the rheological parameters of soil. However, it is not always possible to create conditions of pure relaxation. The dynamometer that records the process of stress relaxation has a certain stiffness, which results in a slight prologing of the relaxation process in time. Hence, the problem appears to be a mixed problem of relaxation and creep in which the specimen is subjected to a special self-loading regimen under a variable damping load. In such cases, for processing of the test results, it is necessary to solve the problem of interaction between a dynamometer of finite stiffness and a soil specimen with rheological properties. Examples of solutions to such problems are discussed in later Chapters.

Field methods of rheological testing of quasi-single-phase soils also do not differ from the standard methods. However, special methods of processing the test results need to be adopted for determining the rheological parameters in the various problems. In this case, the stress-strain state of soil is not homogeneous and it may be determined theoretically.

The field methods of rheological testing (Fig. 3.3) include the following: plate load tests under static loading; plate loading-cum-relaxation tests in which the forces are measured by a dynamometer for a given settlement; pressure and relaxation tests in which dynamometric elements of finite stiffness (plate, round tip etc.) are embedded in the soil.

Rheological tests are much more complicated in the case of quasi-two-phase soils. Therefore, it is advisable to determine the rheological properties

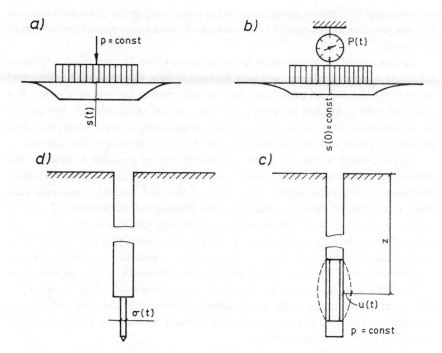

Fig. 3.3. Schematic representation of field methods for testing rheological properties of soils.

a—static plate load test; b—plate loading-cum-relaxation test; c—pressure meter test; d—relaxation test using a dynamometric plate*.

of soil separately under shear and during consolidation. In all cases, shearing tests should be conducted under such loading conditions as do not produce additional volumetric strain and pore pressure. Examples of such conditions are pure shear or simple shear. The effect of consolidating load on shear behaviour of soil can be studied by conducting consolidated undrained tests and measuring the pore pressure. Relaxation tests under shear are also extremely effective because a large number of soil characteristics, including long-term strength, can be determined from these tests.

The volumetric changes in water-saturated clayey soil are accompanied by expulsion of water from the pores. They can be investigated either in consolidation test set-ups with one- or two-way drainage or on set-ups used for testing spherical specimens with central or surface drainage for radial flow of water.

While studying the behaviour of saturated clays in field conditions, only short duration tests should be conducted to eliminate drainage. Otherwise, un-

---

*sic; however, "c" should be "d" and vice versa—Technical Editor.

surmountable difficulties are encountered in processing of the test results because it is not possible to distinguish between time-dependent creep and consolidation processes.

In this regard, relaxation tests are more convenient. During stress relaxation extending over a few hours, soil may be treated as a single-phase medium and the reduced rheological parameters may be determined for soil as a whole. For changing over to the soil skeleton parameters, it suffices to identify that part of the deformation which is occurring due to compression of pore water. It is then possible to relate soil deformation with the effective stresses in the skeleton.

In conclusion it may be mentioned that if it were possible to describe the rheological properties of clayey soil by an equation covering all the rheological processes in soil, independent of the type of test and loading conditions, then there would be no need to determine the rheological parameters of soil by various methods. It would be possible to determine the rheological parameters with sufficient accuracy and reliability merely by conducting one or two tests. Unfortunately, theoretical analysis and suitable equations to meet the above requirements have not been developed to date. However, a few experimental studies have been reported in this direction. It has been established, for example, that creep curves under static and kinematic loading are adequate and that the creep curves obtained from the results of kinematic tests can be readily modified into creep curves for static loading.

4. Equipment and Techniques for Rheological Testing of Clayey Soils

One of the main tasks in rheological testing of clayey soils is the preparation of undisturbed specimens corresponding to the initial state and remoulded specimens with given density and moisture content. This helps to ensure that the stress-strain state of the specimen is as close as possible to that of the actual soil mass. Since the initial stress-strain state of the mass is not known, preloading of the specimen is carried out by the simplest method, such as isotropic compression (on triaxial compression set-up) or consolidation (in consolidation or shear testing set-ups) under a load of intensity $\gamma \cdot h$ (where h is the depth from which the soil specimen was collected). The problem of reproducing the initial stress-strain state is more complex in the case of water-saturated soils because this involves reproduction of not only the stress-strain state of the skeleton, but also the initial pressure of pore water.

It is often necessary to study the effect of total principal stresses or consolidating load on the rheological properties of clayey soils. In such cases the problem is solved in a different manner.

Firstly, the specimen may be loaded in excess of the initial stress by applying a normal consolidating load or producing isotropic compression of varying magnitudes, followed by the application of shearing stresses. If the soil is dense, then in the course of isotropic compression and subsequent deviatory loading, the change in density is insignificant and may be neglected. In this case, all soil characteristics are specified at the initial density and moisture content. If the soil is not sufficiently dense, then the soil density may change significantly. In this case, the soil characteristics would depend not only on the magnitude of isotropic compression, but also on the varying density of the soil skeleton. This factor is particularly important in determining the parameters of the strength envelope because the angle which this line makes with the axis of σ does not represent just the angle of internal friction, but also includes the angle due to hardening.

In such cases, the effect of total principal stresses (or consolidating load) and varying density can be studied separately by conducting the tests described below. The soil specimens are loaded in excess of the initial stress-strain state under the maximum consolidating load and the load is then released by varying degrees. After this step all specimens will be of almost the same density prior to application of deviatory loading; therefore, the effect of consolidating load on mechanical properties of soil can be studied at fixed density.

In engineering practice it is often required to prepare identical specimens of given density and moisture content with remoulded clayey soil. This is necessary in order to obtain specimens whose physical properties are close to the natural state of soil of given density and moisture content used in construction of seepage protection elements of rockfill dams and construction of foundations by replacing weak soils. There are two methods of preparing remoulded specimens of given density and moisture content.

In the first method (standard consolidation) the specimens of given density are subjected to impacts from a falling load moving along a rod attached to the top platen that is mounted freely on the cell containing the soil. With this method, dry soil of density of 1.6 to 1.8 g/cm^3 can be prepared. However, when specimens of greater height (10 to 12 cm) are prepared by this method, non-homogeneity along the height results, which is reflected in the experimental results. Hence it is necessary to satisfy the condition $h < 2d$ and restrict the range of variation of moisture content to $w < w_{opt}$, because at $w > w_{opt}$ this method ceases to be effective.

In the method proposed by the author, identical remoulded soil specimens with given density and moisture content are prepared in the following manner. Air-dried, ground powder is sifted through a sieve of size 0.5 mm. It is spread in the form of a 1–2 cm thick layer in an $80 \times 60 \times 5$ cm trough, wetted uniformly by spraying water over it, and is then thoroughly mixed. This process is repeated until the desired moisture content is achieved uniformly in the whole soil volume. The uniformly moistened powder is kept in an exsiccator for a day or more depending on the percentage of clayey fractions, in order to achieve a more uniform distribution of moisture. Test samples are now taken to confirm that the specimen has the desired moisture content.

The wetted powder is consolidated in a perforated cylinder (Fig. 3.4) placed inside a vacuum chamber with a sliding cover. A rubber packing and a displacement measuring device are mounted on top of the cover. The vacuum produced inside the chamber is measured by means of a vacuum meter. Due to the vacuum, a pressure of almost any magnitude can be transmitted to the perforated loading platen through the sliding cover. The magnitude depends on the ratio between the area of cover F_c and platen area F_p because $\sigma = p_{vac} F_c / F_p$ (where p_{vac} is the vacuum inside the chamber). In the given case a consolidating pressure of 3 MPa was developed at $p_{vac} \approx 0.1$ MPa. The vacuum created in the chamber helps remove air from the moistened powder during consolidation, thereby producing a relatively dense packing. In this manner, specimens with a high degree of water saturation can be obtained with any moisture content—a task difficult to accomplish by any other method.

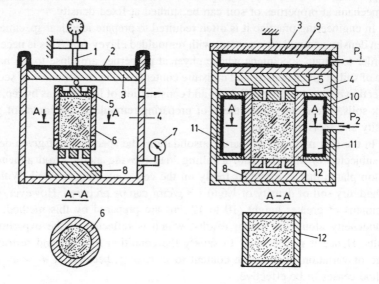

Fig. 3.4. Set-up for preparing clayey soil specimens (a) of given density and moisture content and (b) given stress level,

1—dial indicator; 2—packing; 3—cover; 4—chamber; 5 and 8—perforated platens; 6—perforated cylinder; 7—vacuum meter; 9—rubber chamber; 10—steel casing; 11—lateral compression chamber; 12—steel plates.

If it is required to obtain specimens with water saturation close to unity, then the sliding cover of the chamber should be locked until the requisite vacuum is created in the chamber. This helps in virtually total elimination of air from the moistened powder. When the locking device of the cover is later released, it leads to consolidation of two-phase material, consisting of mineral particles and water, thereby producing a water-saturated specimen.

This method has certain advantages over the well-known method of preliminary consolidation of soil paste. Firstly, the time required for preparation of saturated clayey soil specimens is reduced by tens and hundreds of times. Secondly, it becomes possible to prepare saturated soil specimens of 10–12 cm height, which is impossible by the usual consolidation techniques.

Thus, an efficient and easy method has been developed for acceleration preparation of identical remoulded specimens from clayey soil with given density and moisture content.

The design of the triaxial unsymmetrical compression set-up developed by the author (Fig. 3.1e) permits testing of cubical ($100 \times 100 \times 100$ mm) specimens from clayey soil of given density under compound state of stress ($\sigma_1 \neq \sigma_2 \neq \sigma_3$) for an arbitrary loading trajectory. The set-up has provision for measurement of pore pressure and strain components $\varepsilon_1 \neq \varepsilon_2 \neq \varepsilon_3$ within an accuracy of 10^{-5}. The set-up designed by the author is a modified version of A.L. Kryzhanovskii's triaxial compression set-up and is protected by two proprietary licenses.

The triaxial compression-cum-tension set-up (Fig. 3.1b) developed by the author in collaboration with E.A. Vorob'ev is likewise protected by a proprietary license. The specimen is placed in the chamber of the triaxial compression-cum-tension set-up after filling the chamber with water. When pressure is created in the chamber, the rubber shell is pressed against the soil surface and the specimen concomitantly experiences compressive and tensile stresses. The elongation of the specimen is measured with a dial indicator pressed against the platens. The volumetric strain is obtained by calculating the volume of water released into the chamber with consideration of the specimen elongation. With this set-up, it is possible to test clayey soil specimens of special profiles (in the shape of dumb-bells, solid and hollow cylinders) under conditions of axisymmetrical, triaxial tension and compression, and to determine the rheological parameters of soil when the sum of principal stresses is equal to or less than zero, i.e., in the region of tensile stresses. Depending on its shape, the specimen experiences a compound state of stress that is described by the parameters $\mu_\sigma = 1$ or $\mu_\sigma = -1$.

When a cylindrical specimen is subjected to lateral compression, the state of stress is defined by $\sigma_1 = \sigma_2 = p$ and $\sigma_3 = 0$ (where p is pressure in the chamber). When the side surface is constrained by the rigid split casing, the specimen is subjected to simple tension, i.e., $\sigma_1 = \sigma_2 = 0$; $\sigma_3 = -p(S_{end} - S_{mid})/S_{mid}$ where S_{end} and S_{mid} are the area of cross-section of the end face and the middle section respectively.

Different states of stress are produced when dumb-bell-shaped specimens with different S_{end} to S_{mid} ratios are tested. For example, when $S_{end} = 2S_{mid}$, we have $\sigma_1 = \sigma_2 = p$ and $\sigma_3 = -p$, which represents pure shear in the absence of volumetric strain. At other ratios of S_{end} to S_{mid}, it is found that $\sigma_1 = \sigma_2 = p$ and $\sigma_3 = -p(S_{end} - S_{mid})/S_{mid}$. In all these cases, the Nadai-Lode parameter $\mu_\sigma = 1$. For obtaining other states of stress, the dumb-bell-shaped specimen is subjected to pressure applied simultaneously on the inner and outer surfaces.

The design of the set-up and the method of testing clayey soil specimens are simple. They are within reach of any research laboratory and are widely used in engineering practice [14, 17, 22]. It may be mentioned that the dumb-bell-shaped specimens are made from cylindrical samples by removing the extra material with a wire. The wire is moved along the cylindrical specimen along the guideway attached to the frame. The position of the cylindrical sample is fixed with respect to the frame but can be rotated about its axis.

A number of modified designs based on the triaxial compression-cum-tension set-up have recently been developed. For instance, at the Gidroproekt Research Institute, S.Ya. Zhuk, Yu.K. Zaretskii and E.I. Vorontsov have devised a modified design for developing chamber pressure of up to 3 MPa, which is required for testing compact loams from core walls of high dams in order to investigate crack formation under compound state of stress.

The design developed at Giprotyumen'neftegaz is capable of creating chamber pressure of up to 7 MPa for testing frozen soils.

The triaxial compression set-up developed at the V.V. Kuibyshev Moscow Civil Engineering Institute permits testing of cylindrical soil specimens with provision for measuring pore pressure at the base and in the middle of the specimens. Volumetric strains are measured within an accuracy of 10^{-5} by using high-pressure quartz tubes of small internal diameter. The set-up is equipped with a device for compensation of temperature-induced volumetric changes in the chamber fluid. In addition to the above, the set-up is capable of producing the initial stressed state in the specimen and initial hydrostatic pressure in pore water, corresponding to the initial stress-strain state of the actual soil mass. Subsequent tests are conducted by the closed system, wherein the excess pore pressure produced in the specimen is measured for different trajectories of additional loading.

The ring shearing set-up was developed by the author in collaboration with Yu.S. Grigorev and V.A. Tishchenko and has been granted a proprietary licence. It is meant for testing cylindrical specimens under consolidating load and torque. Depending upon the height of the specimen, it may experience shearing at a particular ring section or shearing due to torque. The advantage of this test is that the shear surface does not change during the test. Furthermore, after the specimen has undergone shearing, it can be additionally twisted in the shear plane to determine the residual strength.

The above set-up can be used for testing standard specimens of cross-sectional area 40 cm^2. Shearing occurs along a strictly fixed plane with constant area of cross-section, which is an important consideration in rheological tests.

The set-ups and devices described above have been fabricated and are being used in research work by the V.V. Kuibyshev Moscow Civil Engineering Institute and other organisations.

Tests conducted for the purpose of determining the rheological parameters of soil should conform to the theoretical principles on which the experiments are based. Therefore, a specific test procedure should be developed or adopted for each particular case involving one or the other theory.

4

Rheological Parameters of Partially Saturated Clayey Soils

1. Parameters of Strength and Hardening

The widely acknowledged Coulomb's strength theory for soil medium is interpreted in various ways. The strength parameters φ and c are often treated as parameters of the strength envelope without relating them to the concepts of friction and cohesion. Thus, by implication they are specified at the initial density and moisture content of soil. This leads to large data scatter whereby φ and c cease to be invariants as far as the state of stress and loading trajectory are concerned.

The author believes that the strength condition formulated by Coulomb is lacking in an additional condition concerning hardening due to progressive shearing strain and slipping of soil particles (aggregates). Actually, Coulomb's strength condition implies that on a given plane ν, the sliding of one set of particles with respect to the other is possible only when $\tau_\nu > \tau_\nu^*$, wherein $\tau_\nu^* = \sigma_\nu \operatorname{tg} \varphi + c$. This formulation is acceptable if the strength parameters φ and c correspond to the initial density of soil skeleton and are used only for estimation of the limit state. However, these parameters are also employed for developing the plastic flow theory and for describing the stress-strain state of soil in the prelimit state. In the latter application of the strength parameters, it is necessary to consider the hardening of soil in the course of deformation and the manner in which the slip planes are oriented in space.

Let us consider shear diagrams of soil for two cases: when the soil is not capable of hardening and when it has this capability (Fig. 4.1). In the first case, the diagram will represent elastoplastic failure while in the second case, it will represent elasto-viscoplastic failure with hardening. In the first case, there is no plastic deformation unit the moment failure occurs. In the second case, each load step produces hardening because the deformation decays with time. This may be attributed to shearing and volumetric deformations, which largely determine the state of soil at the time of failure. This mechanism of development of shearing strain implies that any increment of shearing stresses at constant σ and φ and fixed orientation of slip planes leads to transfer to cohesion after the passage of a certain period of time (Fig. 4.1a). This process continues until the hardening potential is completely exhausted and the soil acquires the state of plastic flow.

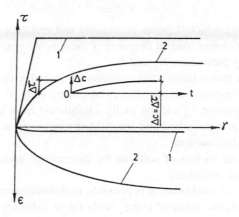

Fig. 4.1. Shearing diagrams for clayey soil.

1—under elastoplastic deformation without hardening; 2—under elasto-viscoplastic deformation with hardening.

Hardening due to volumetric and shearing strains of soil in the process of loading can be taken into account if the $c(\varepsilon)$ and $c(\gamma)$ relations are known. It follows from the foregoing that the overall hardening under purely deviatory loading can be determined through the tangential modulus of shear, i.e.,

$$\Delta c = \Delta \tau = \Delta \gamma G_v^p. \qquad \ldots (4.1)$$

The relations for $c(\varepsilon)$ and $c(\gamma)$ can be represented in the following form:

$$\Delta c(\varepsilon) = K_\varepsilon \Delta \varepsilon^p; \quad \Delta c(\gamma) = K_\gamma \Delta \gamma^p,$$

where K_ε and K_γ are hardening parameters determined experimentally.

Hence, Δc may be expressed as follows:

$$\Delta c = K_\varepsilon \Delta \varepsilon^p + K_\gamma \Delta \gamma^p = G_v^p \Delta \gamma.$$

Two tests need to be conducted for determining K_ε and K_γ: one in which the volumetric deformation is ignored and the other in which it is taken into account. From the first experiment we determine $K_\gamma = \Delta c / \Delta \gamma$ and from the second experiment $K_\varepsilon = (\Delta c - \Delta c_1)/\Delta \varepsilon$.

At the present level of development of the theory of strength for soils, it is necessary to determine the following strength parameters:

— Angle of internal friction φ, which does not depend on the density of dry soil and strain rate, but on the moisture content and the mineralogical and grain size composition of soil.

— Cohesion c, due to rigid crystalline and flexible aqueous colloidal bonds; it depends on density and moisture content and may have the following values:

$$c_{m} > c_{stab} > c_l > c_{res}$$

— Viscosity η, which depends on density and moisture content of soil and the consolidating load; it determines the viscous resistance under shear.

— Hardening parameters K_ε and K_γ.

It is thus seen that cohesion and viscosity of soil play the governing role in moulding the deformation characteristics of clayey soils. It may be mentioned that structural cohesion c_{str} can be easily established from the results of shear tests conducted on undisturbed and remoulded specimens at identical values of density and moisture content.

The strength parameters of soil can be determined under static, kinematic and relaxation test conditions.

Under static load conditions, it is possible to determine not only the strength parameters of partially saturated clayey soils for an arbitrary loading trajectory, but also the parameters that describe the deformability and viscosity of soil. For this, it is necessary either to test three identical specimens on torsion testing set-up or to conduct a torsion test at three levels of consolidating load $\sigma_1 > \sigma_2 > \sigma_3$ following trajectories 6 and 8 (see Fig. 3.2).

Hardening coefficient K_ε may be determined from the envelope of $\tau - \sigma$ curve and curve $\varepsilon - \sigma$ for the load trajectories 1 and 2 respectively (see Fig. 3.2). Since the trajectories of deviatory loading are coincident and the normal stresses are equal, the difference between the limiting values of shearing stress may be attributed solely to the difference in the density of soil skeleton.

At constant density of dry soil, the hardening coefficient of soil K_γ can be determined from the $\gamma - \tau$ curves, subject to the assumption that after each load step the decay of shearing strain is caused by hardening due to shearing strain and not due to change in soil viscosity.

Finally, curves $\gamma(t) - \tau$ can be used to determine the nature of variation of cohesion as a function of shearing strain and the duration for which each load step is applied. This is based on the assumption that there is no volumetric strain and that hardening occurs due only to structural changes in soil and re-orientation of particles etc. It is also assumed that viscosity remains constant. Keeping the foregoing in mind, the equation of viscoplastic flow can be written as follows:

$$\eta^{vp}\dot{\gamma} = \tau - \sigma \, tg \, \varphi - c(\gamma). \qquad \ldots (4.2)$$

It may be noted that $\dot{\gamma} > 0$ at $\tau > \sigma \, tg \, \varphi + c(\gamma)$.

The relation between cohesion and deformation conditions can be written in the form:

$$c(\gamma) = c_0 + K_\gamma \gamma, \qquad \ldots (4.3)$$

where c_0 is cohesion at the beginning of load application; $K_\gamma = (c_\infty - c_0)/\gamma^*$ (here c_∞ is the final value of cohesion, γ is the shear strain varying with time and γ^* its stabilised value for the given load step).

Substituting (4.2) in (4.3), it is found that

$$\frac{d\gamma}{\tau - \sigma\,\mathrm{tg}\,\varphi - c_0 - K_\gamma\gamma} = \frac{dt}{\eta^{vp}}. \qquad \ldots (4.4)$$

The solution to the above equation is:

$$-\frac{1}{K_\gamma}\ln[\tau - \sigma\,\mathrm{tg}\,\varphi - c_0 - K_\gamma\gamma] = \frac{t}{\eta^{vp}} + c.$$

As $\gamma = 0$ at $t = 0$, it follows that

$$c = -\frac{\gamma}{K_\gamma}\ln(\tau - \sigma\,\mathrm{tg}\,\varphi - c_0).$$

Finally, we have

$$\ln(1 - A\gamma) = -Bt, \qquad \ldots (4.5)$$

where

$$A = \frac{K_\gamma}{\sigma\,\mathrm{tg}\,\varphi + c_0}; \quad B = \frac{K_\gamma}{\eta^{vp}}.$$

Relation (4.5) can be written in the form:

$$\gamma = \frac{1}{A}(1 - e^{-Bt}).$$

Upon substituting the above expression in (4.3), it is found that

$$c(\gamma) = c_0 + K_\gamma\frac{1}{A}(1 - e^{-Bt}). \qquad \ldots (4.6)$$

Thus, a relation has been established between time-dependent cohesion and the rheological properties of soil under single-step loading. As the increment of shear strain increases from one load step to the next, while that of cohesion decreases, it is obvious that parameter K_γ will vary and tend to zero at $\tau \to \tau^*$ (where τ^* is ultimate strength of soil).

Results of some tests conducted on clayey soils are given below for stepped loading (Fig. 4.2) at moisture content of 0.1, 0.12 and 0.18 and different values of dry soil density (along trajectories 1, and 2, Fig. 3.2). The angle of internal friction does not depend on density although the latter has a significant effect on cohesion. Both these parameters are significantly dependent on moisture content.

Parameter B is first determined from the creep curves based on formula (4.5). Then the viscosity is found from the relation $\eta^{vp} = K_\gamma/B$. By comparing this value of η^{vp} with its value obtained from the same experiment based on the $\dot{\gamma}-\tau$ relation, it is possible to confirm the validity of the mechanism of viscoplastic deformation described above.

70

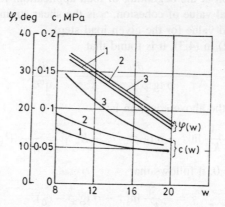

Fig. 4.2. Variation of the parameters of strength of clayey soil with moisture content.
1—$\rho_d = 1.45$ g/cm^2; 2—$\rho_d = 1.5$ g/cm^3; 3—$\rho_d = 1.55$ g/cm^3.

Under kinematic loading it is possible to test clayey soil specimens on shearing set-ups at constant consolidating load and on triaxial set-ups by following the crushing load trajectory (see Fig. 3.2). The advantages of kinematic loading are obvious. In this loading regimen it is possible to test soils in a wide range of variation of strain rate and to determine their rheological parameters in a relatively short period of time.

The $\tau - \lambda$ curves plotted for different values of the displacement velocity (where λ is linear displacement of the head of the loading ram) are the most typical of rheological behaviour. The maximum shearing strength is attained at identical displacements $\lambda^* = $ const., corresponding to deformation $\gamma^* = $ const. When the displacement is large, all the curves lie in the zone or residual strength. The $\tau_{res} - \sigma$ curve passes through the origin of co-ordinates. It is noteworthy that strain rate does not affect the inclination of the strength envelopes i.e., the angle of internal friction does not depend on strain rate. On the other hand, the cohesion or viscous resistance under shear is sensitive to strain rate.

The coefficient of viscoplastic flow $\eta^{vp} = (\tau - \tau^*)/\dot{\gamma}$ can be determined from the $\tau - \dot{\gamma}$ relations (Fig. 4.3) in the stress range $\tau_{max} - \tau_{min}$.

Hence, on the basis of tests conducted under kinematic loading (Figs. 4.4 to 4.6) we first of all determine strength parameters $\tau_m^* > \tau_l^* > \tau_{res}^*$, depending on normal pressure σ for the given moisture content and density of soil. If the density of soil changes in the course of loading, then the shape of the $\tau - \gamma$ curves undergoes a change, i.e., the peak becomes more prominent in the case of consolidated soils and tends to level out in the case of loose soils.

From a comparison of the $\tau - \gamma - t$ curves obtained from tests conducted by M.B. Kornoukhov under stepped and kinematic loading, it can be seen that one

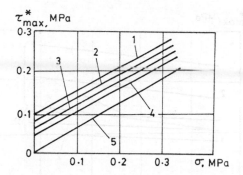

Fig. 4.3. Variation in peak strength τ_{max}^* with σ at different values of strain rate.

1—$\dot{\gamma} = 0.01$ min^{-1}; 2—$\dot{\gamma} = 6.25 \times 10^{-4}$ min^{-1}; 3—$\dot{\gamma} = 3.9 \times 10^{-5}$ min^{-1}; 4—$\dot{\gamma} = 2.44 \times 10^{-6}$ min^{-1}; 5—residual strength at $\dot{\gamma} = 0.01$ min^{-1} (data from M.B. Kornoukhov).

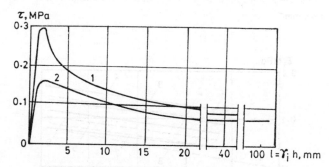

Fig. 4.4. The $\tau - \gamma$ curve plotted on the basis of results of a torsion test on a clay sample taken from a landslide slope ($\gamma_d = 1.31$ g/cm^3; $w = 0.38$) of the same moisture content and density at $\sigma = 0.3$ MPa (data from M.B. Kornoukhov).

1—undisturbed soil; 2—remoulded soil.

curve may be obtained from the other and vice versa. For example, at $\gamma = \dot{\gamma}t$, it is possible to change over to the co-ordinates $t = \gamma/\dot{\gamma}$ under kinematic loading (Fig. 4.7a). A similar conversion is possible with the test results obtained under static loading by changing over first from the $\gamma - t - \tau$ curves to the curves $\dot{\gamma} - t$ and then to the curves $\gamma - t$ (**Fig. 4.7 b**).

Some difficulty is encountered in determination of τ_l^* when it is necessary to describe the $\tau - \gamma$ curve at strain rates of the order of $\dot{\gamma} \times 10^{-6}$ sec^{-1}. As $\gamma^* \approx 0.5$, it would take 0.5×10^5 sec ≈ 6 days to attain γ^*. But at $\dot{\gamma} = 10^{-7}$ sec^{-1} it would take 60 days. Obviously, strain rates of this magnitude are not desirable. It is much easier and more effective to determine the limit

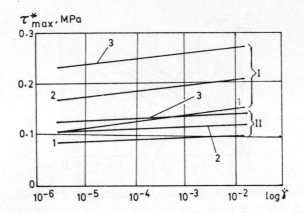

Fig. 4.5. Variation of peak strength with strain rate at different values of consolidation load (data from M.B. Kornoukhov)*

1—σ = 0.1 MPa; 2—σ = 0.2 MPa; 3—σ = 0.3 MPa; I—loess loam soil; II—clay; $\dot{\gamma}$ represents shear strain rate, min^{-1}.

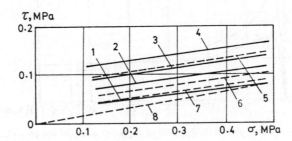

Fig. 4.6. The $\tau_i - \sigma$ curves for different values of shear strain (data from M.B. Kornoukhov).

1—γ = 0.01; 2—γ = 0.05; 3—γ = 0.15; 4—γ = 0.3; 5—γ = 1.1; 6—γ = 3.8; 7—γ = 10; 8—$\gamma \rightarrow \infty$.

of long-term strength by the relaxation test. In this test, the specimen is loaded at a large strain rate to the level corresponding to maximum value of stress τ_{max} and limiting value of shearing strain γ^*. The loading ram-dynamometer-specimen' system is then set in the relaxation mode. In view of the fact that the dynamometer has some stiffness, it is not possible to attain pure relaxation. Nonetheless, the stresses in the specimen and, consequently, the forces recorded by the dynamometer experience relaxation.

---

*In Figure, log should read tg—Technical Editor.

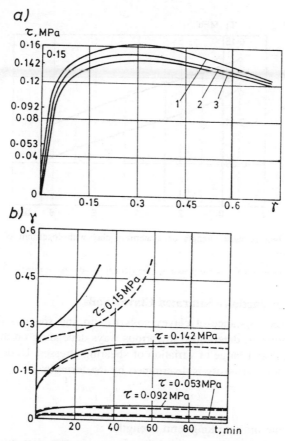

Fig. 4.7. Comparison of rheological curves obtained (a) under kinematic loading and (b) static loading (curves obtained from recomputed data are shown by the dotted line).

In this manner, the curve of long-term strength of the specimen can be determined on the basis of a single test. This technique is similar to the method of testing under uniaxial loading proposed by S.S. Vyalov [1, 17, 22]. The τ_l vs $\ln t$ relation obtained by the above method is linear, i.e., the stresses in the specimen decrease in direct proportion to the logarithm of time (Fig. 4.8).

The variation of long-term strength with time can be obtained in pure form if the dynamometer stiffness is taken into account while solving the problem of stress decay in the 'soil dynamometer' system with restraint on total displacement $\delta = \Delta\gamma(t)h + \Delta l(t) = \text{const}$. (where $\Delta\gamma(t)$ is time-dependent shear strain after clamping of the dynamometer, h is height of specimen and $\Delta l(t)$ is elongation of the dynamometer in the process of relaxation).

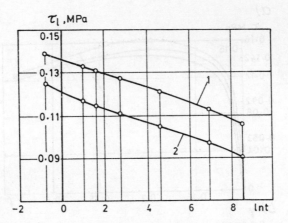

Fig. 4.8. Variation in peak strength of quaternary clay with logarithm of time (data from M.B. Kornoukhov).

1—for dynamometer, 1 kN; 2—for dynamometer, 10 kN; t—duration, min.

2. Viscosity of Partially Saturated Clayey Soils

The viscous resistance of clayey soils to slip and shear is described by a viscosity parameter, determined in various ways depending on the adopted rheological model and range of variation of shearing stresses. Under conditions of viscoelastic flow, viscosity is determined by the formula:

$$\eta^{vp} = (\tau - \tau_l^*)/\dot{\gamma}^{vp}, \qquad \ldots (4.7)$$

where τ^* is the limiting long-term strength.

In the recent past, it became increasingly necessary to determine the viscosity under viscous flow, i.e., at stresses $\tau < \tau^*$. Under these conditions,

$$\eta^v = \tau/\dot{\gamma}^v. \qquad \ldots (4.8)$$

The above necessity stemmed from the associated problem of predicting slow gravitational displacements of shallow landslide-prone slopes interacting with structures of finite stiffness (pipes, bridges, embankment walls etc.).

In addition to the above parameters, it is sometimes required to determine the viscosity under volumetric compression $\eta_\varepsilon^v = \sigma/\dot{\varepsilon}$. A certain relation exists between η_ε^v, η^v and viscosity under linear compression λ^v [see formula (3.1)].

Clayey soil is characterised by three viscosities under shearing and volumetric strain, viz., η^v, η^{vp}, η_ε^v. Parameters η_{sl}^v and η_{sl}^{vp} of viscous and viscoplastic sliding, which are similar to η^v and η^{vp}, can be inducted for describing the viscous resistance of soil in the slip plane. The need to determine these coefficients is also dictated by the fact that it is required to predict the time-dependent processes in soil mass in which the shearing deformation is localised around the slip surface (landslides, sliding wedges etc.).

It was earlier mentioned that the viscosities determined from relations (3.7) and (3.8) are time dependent. This is due, on the one hand, to the discrepancy between the rheological processes and the processes described by eqns. (4.7) and (4.8) and on the other, to the variation with time of rheological parameters, including viscosity and long-term shear strength. In view of the above, eqn. (4.7) would be better written in the following form:

$$\eta^{vp}(t) = [\tau - \tau^*(t)]/\dot{\gamma}^{vp}.$$

However, in this case, the problem becomes more complicated because it is difficult to determine the nature of the $\eta^{vp}(t)$ relation. Therefore, until now, while applying relation (4.7) the damping of deformation was attributed to the variable value of viscosity and it was further assumed that $\tau^*(t) = \tau^* = \text{const}$. The question still remains perplexing. The author believes that the mechanism of damping of shearing strain with time depends less on variable viscosity and more on the variable resistance to shear.

Consider the simplest case, when $c^*(t) = c^* = \text{const}$. and $\eta^{vp}(t) = \eta^{vp} = \text{const}$. The results of tests conducted on clayey soil specimens on a shearing set-up are presented below for the case of stepped loading. The $\dot{\gamma} - \tau$ relations obtained from the experimental data are bilinear for all values of consolidating load (Fig. 4.9). In all the three cases, the point of inflection occurs at the same value of strain rate $\dot{\gamma}^* = \text{const}$. At strain rates $\dot{\gamma} > \dot{\gamma}^*$, the resistance to shearing depends on the residual strength and viscous resistance, i.e.

$$\tau = \sigma \operatorname{tg} \varphi + \dot{\gamma}\eta^{vp}. \qquad \ldots (4.9)$$

At $\tau_l^*(t) = \tau_l^* = \text{const}$. and $\eta^{vp}(t) \neq \text{const}$., the creep curves under static load can be easily described by the following relation:

$$\dot{\gamma} = (\tau - \tau^*)/\eta^{vp}(t). \qquad \ldots (4.10)$$

Various authors have proposed different relations for describing the nature of $\eta^{vp}(t)$ curves:
—according to N.N. Maslov

$$\eta^{vp}(t) = \eta_\infty - (\eta_\infty - \eta_0)e^{\mu t}; \qquad \ldots (4.11)$$

—according to the theory of indefinite creep proposed by Buisman-Pokrovskii

$$\eta^{vp}(t) = \eta_0(1 + \alpha t); \qquad \ldots (4.12)$$

—according to the theory of damping creep

$$\eta^{vp}(t) = \eta_0 e^{\mu t}. \qquad \ldots (4.13)$$

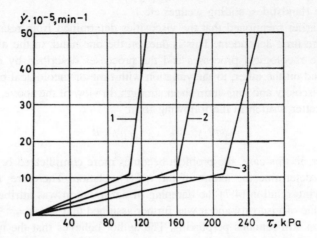

Fig. 4.9. Rheological curves (τ vs $\dot{\gamma}$) for surficial loam from a landslide slope at different consolidating loads.

1—$\sigma = 200$ kPa; 2—$\sigma = 300$ kPa; 3—$\sigma = 400$ kPa.

Upon substituting the above equation in the initial expression (4.10) and integrating, it is found that:

$$\gamma(t) = (\tau - \tau^*)\left(\frac{t}{\eta_\infty} + \frac{1}{\mu\eta_\infty} \ln\left[\frac{\eta_\infty - (\eta_\infty - \eta_0)e^{-\mu t}}{\eta_0}\right]\right);\quad \dots (4.14)$$

$$\gamma(t) = \frac{(\tau - \tau^*)}{\eta_0\alpha} \ln(1 + \alpha t); \qquad \dots (4.15)$$

$$\gamma(t) = \frac{\tau - \tau^*}{\eta_0\mu}(1 - e^{-\mu t}). \qquad \dots (4.16)$$

The first case represents undamped creep which ends with flow at constant strain rate $\dot{\gamma} = (\tau - \tau^*)/\eta_\infty$ because of the inequality $\tau > \tau^*$. In the second case, the strain rate decays with time, although the shearing strain increases proportionately with logarithm of time. The third case represents damping of both strain rate and deformation with time because the viscosity tends to infinity.

Any of the three laws of variation of viscosity described above can be adopted depending on the type of soil and the range of variation of density, moisture content and stresses. The experimental results can be processed accordingly to determine the parameters included in the expressions. For instance, after determining η_0 and η_∞ at time t_0 and t_∞ and strain rates $\dot{\gamma}_0$ and $\dot{\gamma}_\infty$, it can be found that:

$$\mu = \frac{1}{t} \ln \frac{\eta_\infty - \eta_0}{t_\infty - t_0}. \qquad \dots (4.17)$$

Parameters η_0 and α in eqn. (4.15) can be determined as follows. From the initial segment of the $\dot{\gamma}-t$ curve, the value of η_0 is determined for the particular strain rate $\dot{\gamma}_0$. The γ vs $\ln(t)$ curve is now plotted and from the inclination of the line the value of α is determined by the relation

$$\alpha = \frac{\gamma_2 - \gamma_1}{\ln(t_2/t_1)}. \qquad \dots (4.18)$$

Finally, parameters η_0 and μ in eqn. (4.16) are readily found at the initial and final values $\dot{\gamma}_0$ and $\dot{\gamma}_\infty$ of the conditionally stabilised strain rates.

Let us now discuss the case when the viscosity of soil can be considered constant and damping of shearing strain with time occurs due to hardening of soil skeleton due to an increase in cohesion. For this case, eqn. (4.10) can be written as follows:

$$\dot{\gamma}(t) = [\tau - \sigma \cdot \mathrm{tg}\,\varphi - c(t)]/\eta^{vp}(\sigma), \qquad \dots (4.19)$$

where $c(t)$ is time-dependent cohesion and $\eta^{vp}(\sigma)$ is viscosity that depends on normal stress σ.

It is obvious that $\dot{\gamma} = 0$ at $\tau < \sigma \cdot \mathrm{tg}\,\varphi + c(t)$ and $\dot{\gamma} > 0$ at $\tau > \sigma \cdot \mathrm{tg}\,\varphi + c(t)$. In the simplest case, let it be assumed that after the application of a particular load step the variation of cohesion is described by the following relation:

$$c(t) = c_\infty - (c_\infty - c_0)e^{-\mu t}, \qquad \dots (4.20)$$

where $c_\infty = \tau - \sigma \cdot \mathrm{tg}\,\varphi$. The rate of shearing strain due to the given load step will be:

$$\dot{\gamma}(t) = [(c_\infty - c_0)e^{-\mu t}]/\eta^{vp}(\sigma). \qquad \dots (4.21)$$

Upon integrating the above equation, it is found that:

$$\dot{\gamma}(t) = \frac{c_\infty - c_0}{\eta^{vp}\mu}(1 - e^{\mu t}). \qquad \dots (4.22)$$

Suppose the variation of cohesion after application of a particular load step is described by the following relation:

$$c(t) = c_0 e^{(\lambda_1 - \lambda_2)t}, \qquad \dots (4.23)$$

where λ_1 and λ_2 are parameters that describe the rate of appearance and disappearance of new bonds in soil respectively; at $\lambda_1 > \lambda_2$ hardening takes place, while at $\lambda_1 < \lambda_2$ the soil becomes weaker.

After substituting (4.23) in formula (4.21) and integrating, it is found that

$$\dot{\gamma}(t) = \frac{\tau - \sigma \cdot \mathrm{tg}\,\varphi}{\eta^{vp}}t - \frac{c_0}{\eta^{vp}(\lambda_1 - \lambda_2)}[1 - e^{(\lambda_1 - \lambda_2)t}]. \qquad \dots (4.24)$$

The condition $\lambda_1 > \lambda_2$ describes damping creep, $\lambda_1 = \lambda_2$ stable creep and $\lambda_1 < \lambda_2$ progressive creep. It is not too difficult to determine λ_1 and λ_2

experimentally but the difference $\lambda_1 - \lambda_2$ can also be determined from the $\gamma(t)$ curve.

The theoretical basis and the method of determining the viscosity parameters of clayey soils have been described above. The same approach may be adopted for determining the viscosity parameters for the processes of viscous and viscoplastic sliding over a fixed surface.

It is obvious that a certain relation exists between the coefficients of viscosity under sliding and shearing. This relation can be derived by changing over from shearing strains to displacements (see Fig. 2.10), i.e., by substituting,

$$\eta_{sl} h = \eta_{v'} \qquad \dots (4.25)$$

where h is the thickness of the structural element.

By comparing the η_v and η_{sl} viscosity values for clayey soil determined on a direct shearing and torsion set-up, it was found that the thickness of the structural element was $h = 1.66$ m. This is in conformity with the observed results pertaining to shearing strain of clay.

3. Parameters of Deformability of Partially Saturated Soils

The parameters of soil deformability with reference to rheology are G_m, G_l and K_m, K_l which describe the stipulated instantaneous and stipulated stable shearing and volumetric strains of the soil skeleton. In the course of creep, the deformability parameters vary with time in a particular manner, depending on the density and moisture content, the magnitude of working stresses etc. It is natural that the parameters of soil deformability should be included in the equations of state. For example, for a viscoelastic medium, we have

$$\gamma = \tau/G_m + \tau t/\eta^v; \quad \varepsilon = \sigma/K_m + \sigma t/\eta_v. \qquad \dots (4.26)$$

If η^v/t and η_v/t are replaced by $G_l(t)$ and $K_l(t)$ respectively in the above equations, then it is found that

$$\gamma = \tau[1/G_m + 1/G_l(t)]; \quad \varepsilon = \sigma[1/K_m + 1/K_l(t)]. \qquad \dots (4.27)$$

Now suppose that viscosity of soil changes with time as follows:

$$\eta^v(t) = \eta_0^v(1 + \alpha t); \quad \eta_v = \eta_{v_0}(1 + \beta t). \qquad \dots (4.28)$$

Keeping the above in mind, eqn. (4.27) can be written as:

$$\gamma = \frac{\tau}{G_m} + \frac{\tau}{\eta_0^v \alpha} \ln(1 + \alpha t); \quad \varepsilon = \frac{\sigma}{K_m} + \frac{\sigma}{\eta_{v_0} \beta} \times \ln(1 + \beta t). \qquad \dots (4.29)$$

Introducing the notations

$$G_l(t) = \eta_0^v \alpha / \ln(1 + \alpha t); \quad K_l(t) = \eta_{v_0} \beta / \ln(1 + \beta t), \qquad \dots (4.30)$$

eqn. (4.29) can be written in the following form:

$$\gamma = \tau/G(t); \quad \varepsilon = \sigma/K(t), \qquad \dots (4.31)$$

where $1/G(t) = 1/G_m + 1/G_l(t); \quad 1/K(t) = 1/K_m + 1/K_l(t)$.

It is clear from the foregoing that for a viscoelastic medium the moduli of long-term strains are related with viscosity. They may be determined on the basis of the latter or directly from the values of strains $\gamma(t)$ and $\varepsilon(t)$ recorded at different time instants. Obviously, for an ideally viscous medium the moduli of long-term strain will tend to zero with the passage of time. In the case of variable viscosity this will occur after a very long period of time.

When the soil experiences viscoplastic deformation

$$\dot{\gamma} = (\tau - \tau^*)\eta^{vp}; \quad \dot{\varepsilon} = \sigma/\eta_v. \qquad \dots (4.32)$$

In the case of constant η^{vp} and η_v

$$\gamma(t) = \tau/G_m + (\tau - \tau^*)t/\eta^{vp}; \quad \varepsilon(t) = \sigma/K_m + \sigma t/\eta_v. \qquad \dots (4.33)$$

Let us introduce the notation

$$G_l(t) = \frac{\eta^{vp}}{t(1 - \tau^*/\tau)}, \qquad \dots (4.34)$$

as a result of which

$$\gamma(t) = \tau[1/G_m + 1/G_l(t)]. \qquad \dots (4.35)$$

Let us consider relatively complex cases in which the varaiation of viscosity with time is described by the relation:

$$\eta^{vp} = \eta_0^{vp}(1 + \alpha t), \qquad \dots (4.36)$$

then

$$\left.\begin{aligned}
\gamma(t) &= \frac{\tau}{G_m} + \frac{\tau - \tau^*}{\eta_0^{vp}\alpha}\ln(1 + \alpha t); \\
\varepsilon(t) &= \frac{\sigma}{K_m} + \frac{\sigma}{\eta_{v_0}\beta}\ln(1 + \beta t).
\end{aligned}\right\} \qquad \dots (4.37)$$

In this case,

$$\left.\begin{aligned}
G_l(t) &= \frac{\eta_0^{vp}\alpha}{(1 - \tau^*/\tau)\ln(1 + \alpha t)}; \\
K_l(t) &= \frac{\eta_{v_0}\beta}{\ln(1 + \beta t)}.
\end{aligned}\right\} \qquad \dots (4.38)$$

It is evident that $G_l(t) = \infty$ at $\tau \leq \tau^*$, i.e., the rate of viscoplastic strain is equal to zero.

In the equation of viscoplastic flow discussed above, the modulus of shear is determined as a function of viscosity, ratio τ^*/τ and time. This is typical of creep in clayey soils, particularly those with variable viscosity.

The procedure for determining the deformability parameters in eqns. (4.33), (4.35) and (4.37) is the same as in the previous cases. The modulli G_m and K_m are determined from stipulated instantaneous strain by the following relations:

$$G_m = \tau/\gamma_m; \quad K_m = \sigma/\varepsilon_m. \qquad \ldots (4.39)$$

Similarly G_l and K_l are determined from the stipulated stabilised strains as follows:

$$G_l = \tau[\gamma_l(t_{\text{stab}}) - \gamma_m]; \quad K_l = \sigma[\varepsilon_l(t_{\text{stab}}) - \varepsilon_m], \qquad \ldots (4.40)$$

where t_{stab} is the duration of stipulated stabilisation.

The initial values of viscosities η_0^{vp} and η_{v_0} are determined from the initial stipulated stable values of strain rates $\dot{\gamma}_m$ and $\dot{\varepsilon}_m$:

$$\eta_0^{vp} = \dot{\gamma}_m/(\tau - \tau^*); \quad \eta_{v_0} = \dot{\varepsilon}_m/\sigma. \qquad \ldots (4.41)$$

Parameters α and β in formulae (4.37) may be determined on the basis of viscosity at the time instant t_{stab}:

$$\left.\begin{array}{l} \eta^{vp}(t_{\text{stab}}) = \eta_0^{vp}(1 + \alpha t_{\text{stab}}) = (\tau - \tau^*)/\dot{\gamma}(t_{\text{stab}}); \\ \eta_v(t_{\text{stab}}) = \eta_{v_0}(1 + \beta t_{\text{stab}}) = \sigma/\varepsilon(t_{\text{stab}}), \end{array}\right\} \qquad \ldots (4.42)$$

or

$$\left.\begin{array}{l} \alpha = \left[\dfrac{\tau - \tau^*}{\dot{\gamma}(t_{\text{stab}})\eta_0^{vp}} - 1\right]\dfrac{1}{t_{\text{stab}}}; \\[3mm] \beta = \left[\dfrac{\sigma}{\dot{\varepsilon}(t_{\text{stab}})\eta_{v_0}} - 1\right]\dfrac{1}{t_{\text{stab}}}. \end{array}\right\} \qquad \ldots (4.43)$$

In concluding, we shall discuss one more type of rheological equation, examined earlier in Chapter 2. These are equations of hereditary creep [see relation (2.26)] in which the deformation parameters include instantaneous strain moduli G_m and K_m, long-term strain moduli G_l and K_l, creep parameters δ_γ and δ_v, and parameters a_γ, a_ε and τ_1. The first four parameters can be readily determined from the stipulated instantaneous and stipulated stable strain values with the help of formulae (4.39) and (4.40). Creep parameters δ_γ and δ_v can also be determined from creep curves $\gamma(t)$ and $\varepsilon(t)$. Ageing parameters a_γ, a_ε and τ_1 are determined from $\gamma(t)$ vs $\ln(t)$ and $\varepsilon(t)$ vs $\ln(t)$ curves because, in this co-ordinate system, the curves invariably become straight lines (see Figs. 2.14 and 2.15).

$$a_\gamma = \frac{\gamma(t_2) - \gamma(t_1)}{\tau \ln(t_2/t_1)}; \quad a_\varepsilon = \frac{\varepsilon(t_2) - \varepsilon(t_1)}{\sigma \ln(t_2/t_1)}, \qquad \ldots (4.44)$$

wherein $t_2 > t_1 > t_{\text{stab}}$.

Parameter τ_1 may be determined from the $\gamma(t)$ vs $\ln(t)$ or $\varepsilon(t)$ vs $\ln(t)$ curve for the point at which these straight lines intersect the axis of t.

Thus all the parameters of eqn. (2.26) can readily be determined from the results of shear and isotropic compression tests on partially saturated soil specimens under static loading. These parameters can also be determined for other loading conditions by processing the experimental data in an appropriate manner and modifying eqn. (2.26) in conformity with the experimental conditions. For instance, for relaxation testing under shear, eqn. (2.26) must be solved with respect to stresses. However, since a dynamometer of finite stiffness is used in relaxation tests, it becomes necessary to solve the relaxation problem for the system 'soil-dynamometer', taking the dynamometer stiffness into account.

When soil is subjected to constrained compression, it becomes necessary to determine the following parameters: m_{v1}, m_{v2}, m_{v3}, δ_{com} and τ_1. As in the previous cases, m_{v1} and m_{v2} are determined from the stipulated instantaneous (ε_m) and stipulated stable ($\varepsilon_{\text{stab}}$) values of strain, i.e.,

$$m_{v1} = \varepsilon_m/\sigma_1; \quad m_{v2} = (\varepsilon_{\text{stab}} - \varepsilon_m)/\sigma_1.$$

Parameters m_{v3} and τ_1 are found from the $\varepsilon_1(t)/\sigma_1$ vs $\ln(t)$ curve as follows:

$$a_{\text{com}} = \frac{\varepsilon_1(t_2) - \varepsilon_1(t_1)}{\sigma_1 \ln(t_2/t_1)}, \qquad \ldots \ (4.45)$$

where $t_2 > t_1 > t_0$.

The value of τ_1 is determined as the point of intersection of the straight line $\varepsilon_1(t)/\sigma_1$ with the axis of $\ln(t)$ (Fig. 4.10).

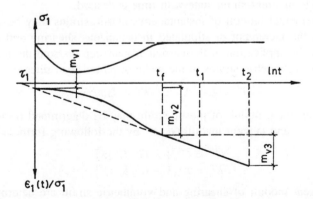

Fig. 4.10. Creep curve under constrained compression for determining the rheological parameters of soil skeleton m_{v1}, m_{v2}, m_{v3} and τ_1 (t_f is the time of completion of consolidation seepage).

4. Parameters of Viscoplastic Flow of Hardening Partially Saturated Clayey Soils

The theory of viscoplastic flow of hardening clayey soils described in Section 2.4 makes it possible to describe the creep process under stepped loading. This has great practical significance: firstly, stepped loading of the foundation is customary in most of the cases encountered in engineering practice and secondly, the existing methods of numerical solutions to engineering problems can take into account more factors for the case of stepped loading than for single-stage loading.

An analysis of the above theory and the rheological equation of state revealed that the rheological model of soil for stepped loading allows many such factors to be taken into acount which cannot be considered in the case of single-stage loading, namely, hardening, loss of strength, change of viscosity, modulus of deformation from one step to the next etc. On the other hand, when these factors are taken into account, complex relations result, thereby enjoining that the numerous parameters be determined at each stage of loading. Given this, such an approach may be justified only for solving engineering problems arising from the construction of unique and huge structures. Thus, according to the theory of viscoplastic flow for hardening clayey soil, at each load step it is necessary to determine the incremental moduli of stipulated instantaneous and stipulated stable shearing and volumetric strains, i.e., G^e, G^p, K^e, and K^p. It may be mentioned that these parameters depend upon the initial state of stress, loading trajectory, parameters of soil strength φ and $c(t)$, hardening parameters K_ε, K_γ, K_t, δ and λ and viscosity parameters $\eta^{vp}(t, \sigma)$.

Generally speaking, it suffices to know the deformation parameters after each load step for evaluating the stabilised stress-strain state of soil mass. However, it is necessary to determine the hardening and viscosity parameters if prediction of the change in stress-strain state with time is desired.

The incremental moduli of instantaneous elastic strains may be determined either from the increment of stipulated instantaneous shearing and volumetric strains after the application of the particular load step or from the increment of shearing and volumetric strains in the wake of partial load release, i.e.,

$$G^e = \Delta\tau/\Delta\gamma^2; \quad K^e = \Delta\sigma/\Delta\varepsilon^e. \qquad \ldots (4.46)$$

The incremental moduli of plastic strain may be determined through elastic moduli G^e, K^e and tangent moduli G^t, K^t by the following formulae:

$$\left. \begin{array}{l} G^p = G^t(1 - G^t/G^e); \\ K^p = K^t(1 - K^t/K^e). \end{array} \right\} \qquad \ldots (4.47)$$

The tangent moduli of shearing and volumetric strain are determined from the total stipulated stable strains by the relations:

$$G^t = \Delta\tau/[\Delta\gamma(t^*)]; \quad K^t = \Delta\sigma/[\Delta\varepsilon(t^*)], \qquad \ldots (4.48)$$

where t^* is the period of stipulated stabilisation of shearing and volumetric strains.

The investigations conducted by the author in collaboration with M.V. Proshin revealed that the incremental moduli of shearing and volumetric strains are significantly dependent on the working stresses and the trajectory of additional loading (Figs. 4.11 and 4.12). Therefore, the directions of the load vectors that produced exclusively elastic strains were fixed. For instance, unloading takes place, i.e., there is no plastic strain, when the load vector is directed within the angle $\pi/2$ (Fig. 4.13). If the load vector is directed outside, then it produces additional loading and progressive increase of distortion and volumetric strain; when the load vector is perpendicular to envelope 7, it represents maximum shearing plastic strain and when it is parallel to envelope 7, it represents maximum volumetric plastic strain. Thus a deformation anisotropy is observed, both in respect of shape as well as volume.

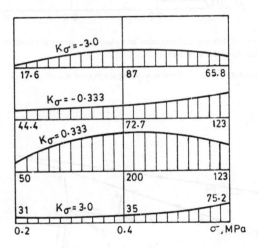

Fig. 4.11. Variation of incremental modulus of shear G^P with σ and τ for partially saturated loam (data from M.V. Proshin).

It may be concluded from the foregoing that increments of plastic strain cannot be related to the potential of the load surface; therefore, it is impossible to determine these increments on the basis of associated plastic flow rule. At every point in the $\tau - \sigma$ plane the increment of the shearing and volumetric plastic strains depends on the initial state of stress and the direction of the loading trajectory. In other words, the increments of shearing and volumetric plastic strains are described by extremely complex relations of the type:

$$\left.\begin{array}{l} \Delta\gamma^P = \Delta\gamma^P(\sigma,\tau,K_\sigma,\mu_\sigma,t^*); \\ \Delta\varepsilon^P = \Delta\varepsilon^P(\sigma_1,\tau,K_\sigma,\mu_\sigma,t^*). \end{array}\right\} \qquad \ldots \text{ (4.49)}$$

84

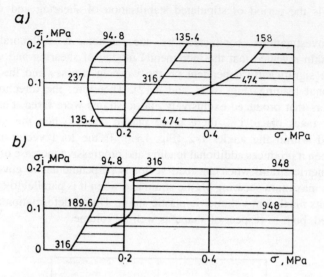

Fig. 4.12. Variation of incremental modulus of volumetric strain K^p with σ, ε and $K_σ$ for partially saturated loam (data from M.V. Proshin).

a—for loading trajectory $K_σ = 3$; b—for loading trajectory $K_σ = -3$.

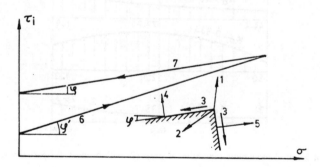

Fig. 4.13. Schematic representation of the growth of plastic strain depending on the direction of load vector.

1—loading; 2—unloading; 3—partial loading; 4—direction of maximum shearing strain; 5—direction of maximum volumetric strain; 6 and 7 : envelopes without and with precompression of soil respectively.

The variation of G^p with the above parameters has been discussed in the theory of plastic flow described earlier (see section 2.4).

It follows from the foregoing that for determining the stress-strain state of soil mass after each load step, the variation of the parameters must be known not only

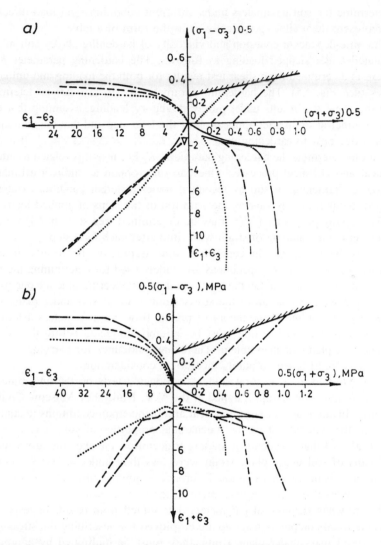

Fig. 4.14. Results of triaxial testing of clayey soil (data from N.M. Mkrtchyan).

a—under axial symmetry condition ($\varphi = 11.3°$, $c = 0.25$); b—under plane strain condition ($\varphi = 18.4°$, $c = 0.26$).

as a function of the stress state attained, but also the direction of the loading trajectory. As this direction differs at each point of the non-homogeneously stressed soil mass and is not known, the load direction of the previous step is adopted as the initial direction for the current step. Hence it becomes necessary to determine the soil parameters under different conditions and load directions and represent them either graphically or in the form of a table.

The time-dependent cohesion and viscosity of hardening clayey soil may be calculated under stepped loading as follows. The hardening parameter $K_\varepsilon = \Delta c / \Delta \varepsilon$ is determined from shear test results for both the loading and unloading curves (see Fig. 1.5). The hardening parameter $K_\gamma = \Delta c / \Delta \gamma$ is determined from the shear test results under purely deviatory loading, assuming that there is no volumetric strain after application of the load step and that the damping of strain is due only to hardening under shear, i.e., $K_\gamma = \Delta \tau / \Delta \gamma \approx c^t$. It is more difficult to determine the hardening parameter K_t because it is related to internal physical and chemical processes. One can only obtain an indirect estimate of the rate of hardening of soil by testing it under different conditions (rates) of stepped loading, i.e., by varying the duration of the steps of applied loads.

The varying parameter $\eta^{vp}(t)$ may be determined (see section 4.3) from the curves depicting damping of strain with time after each load step.

It may be mentioned in conclusion that at present the results of three-dimensional testing of soil specimens are widely used for determining the rheological parameters of soil. On the other hand, in engineering practice one mostly has to deal with evaluation of the stress-strain state of soil under plane strain conditions. According to the theory of plastic flow, the parameters determined from three-dimensional testing must be suitably corrected before they can be applied to a plane strain problem. There are guidelines for carrying out such corrections [5, 6] based on purely theoretical considerations.

N.M. Mkrtchyan, a doctoral student, conducted tests on the unsymmetrical triaxial compression set-up designed at the V.V. Kuibyshev Moscow Civil Engineering Institute under axisymmetrical and plane strain conditions to study the effect of the type of test on the mechanical properties of soil. It can be seen from Fig. 4.14 that for identical loading trajectories, the shearing and volumetric strains of soil under plane strain were less than under an axial symmetry condition. It is noteworthy that the difference is quite significant in terms of the angle of internal friction but very small in terms of cohesion.

Hence, when rheological parameters determined from results of tests on an unsymmetrical compression set-up are employed for describing the stress-strain state of soil mass under plane strain, they must be multiplied by appropriate correction factors.

5

Rheological Parameters of Saturated Clayey Soils

1. Special Features of Rheological Behaviour of Saturated Soils

To solve applied problems, it is desirable to differentiate between three groups of rheological parameters of saturated clayey soils. The first group includes rheological parameters of soil skeleton that describe the growth of shearing strain; the second group includes rheological parameters of soil skeleton that describe volumetric strain and the strain due to constrained compression; the third group includes rheological parameters of pore water that describe volumetric strain and viscous resistance of water to expulsion.

The method of determining the parameters of shearing creep of the skeleton of saturated soil (first group) differs from the method for partially saturated soil only in that the pore pressure must be measured during the test. This makes it possible to change over from total normal stresses to normal stresses in the soil skeleton. In some cases, to avoid build-up of residual pore pressure, the tests are conducted according to the consolidated undrained scheme, which yields good results. Nevertheless, even at constant normal load the shearing stresses sometimes produce volumetric strain due to dilatation, which, in turn, results in pore pressure. Therefore, while testing saturated clayey soils it is desirable to monitor, as far as possible, the pore pressure or, at least, the volumetric strain.

In speaking of the rheological parameters of soil skeleton, one implies the rheological properties of soil skeleton that do not depend on the extent of filling of the pores with water. Such a condition of the skeleton is encountered when the moisture content is in excess of the plastic limit. Otherwise, it is not proper to draw a parallel between the properties of skeleton in saturated and unsaturated states and the test results for partially saturated soil cannot be applied to saturated soil.

It is necessary to carry out consolidation to determine the second group of rheological parameters of the skeleton of saturated clayey soil, because the skeleton displays rheological behaviour only under compression. In this case also it is possible to conduct the tests by a number of different methods.

For example, in a closed system (absence of drainage) it is possible to determine the parameters or volumetric creep of skeleton as well as the volumetric strain of pore water with dissolved gases. Obviously, in the case of fully saturated soil, it is meaningless to conduct such tests. Besides, a partially saturated specimen acquires the state of full saturation under the influence of normal and shearing stresses and it becomes impossible to load it further. As a result, the rheological parameters are determined only in a narrow range of variation of the skeleton stresses and volumetric strain. This is the main drawback of the tests conducted by the closed scheme. However, in some cases such tests are essential. For example, to estimate the pore pressure in clayey foundations and cores of high dams, it is mandatory that the initial pore pressure be determined without taking into account the seepage processes. In view of the foregoing, special attention is paid here to the methods of testing under closed conditions.

The testing of saturated soils under conditions of free drainage is essential in determining the rheological parameters in a wide range of variation of skeleton stresses, density and moisture content of the soil.

Finally, the third group of rheological parameters of clayey soil includes the coefficient of permeability and the modulus of volumetric compressibility of pore water with dissolved gases, which determine the time-dependent process.

Attention is drawn to one more important factor. In engineering practice, one may face a situation in which it is necessary to determine the stress-strain state of saturated clayey soil mass under short-duration external load. In such cases, the formation of and changes in the stress-strain state of the soil mass may be examined for the soil as a whole, i.e., through the total stresses. The reduced moduli of deformation G_{red} and K_{red}, strength parameters C_{red} and φ_{red} and viscosity η_{red} may be adopted for defining the deformability and strength. In this case, all tests should be conducted by the closed system and there is no need to measure pore pressure during the tests.

2. Shear Parameters of Saturated Clayey Soils

To determine the parameters of shear creep of saturated clayey soil (parameters of strength, viscosity and deformability), the tests should be conducted according to the following two schemes: consolidated-undrained and unconsolidated-undrained.

Consolidated-undrained test

These tests are conducted after preliminary consolidation of saturated soil when the full consolidating load has been transmitted to the skeleton. This helps in two ways: the soil can be tested under different conditions of density and moisture content and at various values of normal stresses in the skeleton. However, application of shear after preliminary consolidation of soil often produces additional volumetric strain and, consequently, pore pressure. This factor can be taken into account only by measuring pore pressure in the soil specimen being tested.

For instance, while determining the parameters of strength, the results of triaxial tests should be represented in the form of limiting Mohr's circles for effective stresses. The outcome of this step will be that the limiting Mohr's circles for total stresses will be shifted to the left by a distance equal to the increment of pore pressure. The Mohr envelope will also shift its position in the $\tau - \sigma$ plane, depending on the stresses for which the Mohr's circles are plotted.

It is apparent that the results of the tests conducted on torsion and shearing set up by the consolidated-undrained scheme should also be represented in the τ vs σ_s co-ordinates. It is necessary to ensure absence of drainage during the tests and make provision for measurement of pore pressure. It is difficult to meet these conditions on a shearing set-up because it is impossible to produce a closed system, while measurement of pore pressure on the shear plane is beset with great difficulties. In view of this, shear tests should be conducted under conditions that ensure quick shearing in order to achieve at least partial absence of drainage.

The long-term strength τ_m and τ_l^* and viscosity η^{vp} and η^v of saturated soils are determined in the same manner as was described in the previous chapter for partially saturated soils. The only difference is that these parameters may be attributed to the soil skeleton or the soil as a whole, depending on the stresses (total or effective) for which the results are processed and the analytical scheme which is employed for evaluating the stress-strain state of the soil mass.

As pore pressure is produced during the tests under deviatory loading, it can be taken into account by carrying out direct measurements in the course of the test.

Further, not all the parameters of long-term strength and viscosity are functions of normal stresses. The tests conducted by the consolidated-undrained scheme should help in separately studying the effect of density/moisture content and normal stresses on the parameters of long-term strength of soil. It is wrong to equate the effect of normal stresses on rheological parameters of soil with that of the normal stress induced variations in density of soil skeleton. In order to take these factors into account separately, the tests based on consolidated-undrained method should be conducted according to the two schemes described in Chapter 2.

Unconsolidated-undrained test

These tests are conducted when it is difficult to provide drainage during the process of shearing. This is observed at low values of the coefficient of permeability and when forces are applied rapidly on the soil mass. Under these conditions the parameters of strength and deformability can be determined in terms of both effective and total stresses.

The most difficult aspect of the above method of determining the strength and deformability parameters is that the testing of saturated clayey soils under unconsolidated-undrained conditions often leads to a situation in which the applied normal stresses are taken up by the pore water after a certain degree of consolidation and saturation of the specimen. Under these conditions the shear resistance is governed mainly by the cohesion of soil and all the limiting Mohr's circles in terms of total stresses have the same diameter, irrespective of the magnitude of the intermediate principal stresses $\sigma_2 = \sigma_3$. When pore pressure is taken into account, the Mohr's circles for effective stresses are found to be practically superimposed one upon the other, thereby making it impossible to plot the envelope in the $\tau - \sigma$ plane.

The parameters of long-term strength and viscosity of saturated soils may be determined by unconsolidated-undrained tests either for the skeleton or the soil as a whole. However, there would be significant differences between the two because, in this case (unlike the consolidated-undrained test), the shearing creep will be affected by pore pressure, which can become quite large during the precompression stage. Besides, the reduction (or increase) of strength is not the sole cause of the variation of shear strength with time. The latter is also affected by pore pressure, which in the unconsolidated-undrained tests increases with time due to the delay in volumetric strain.

Hence, if the rheological parameters of soil have been determined in terms of total stresses, they must not be used under any circumstances in computations of the stress-strain state of soil mass based on effective stresses and vice versa. It may be noted that while referring to the stress-strain state of soil mass, the absence of drainage is automatically implied although it may not be specifically stated as such.

3. Parameters of Saturated Soils during Consolidation

A brief description of the process of consolidation of soil mass is necessary before undertaking a description of the method of determination of the rheological properties of saturated clayey soil during consolidation.

The consolidation of saturated soil is related to the interaction between soil phases (solid and liquid) and the changes that occur in their relative mass proportions in a unit volume, both in space and time. The intensity and direction of consolidation of saturated clayey soils is affected most by the drainage conditions, degree of saturation and rheological properties of the soil skeleton and the pore water with dissolved gases. It is also affected by the shape and size of the soil mass being subjected to consolidation. It is thus clear that consolidation is an extremely complex process which depends on a number of factors, involving n parameters. One possible method for solving this problem could be factor analysis to identify the principal and governing consolidation parameters. However, it is preferable to adopt a simpler approach.

After the load has been applied, the process of consolidation can be divided into three stages: initial, intermediate and final.

The *initial stage* of consolidation is described by such a stress-strain state of soil in which consolidation due to flow of water is negligible. Therefore, this stage may be described through total stresses, using the reduced soil parameters G_{red} and K_{red} that represent the behaviour of soil as a whole. The initial values of effective stresses and pore pressure can also be estimated on the basis of initial total stresses. The initial pore pressure is required for assessing the stress-strain state of soil mass during the intermediate stage of consolidation. Thus, the following parameters are required for estimation of the initial stress-strain state of soil: reduced soil parameters G_{red} and K_{red} and pore water compressibility parameter K_w. It may be mentioned that, in general, G_{red} and K_{red} are time-dependent parameters and may, therefore, be represented in the form:

$$\left.\begin{array}{l} 1/G_{red} = 1/G_{red}^e + (1/G_{red}^v)f_\gamma(t); \\ 1/K_{red} = 1/K_{red}^e + (1/K_{red}^v)f_v(t), \end{array}\right\} \qquad \ldots (5.1)$$

where $f_\gamma(t)$ and $f_v(t)$ are time-dependent functions that determine the development of shearing and volumetric strain in saturated soil with time in the absence of drainage.

The effect of time on the initial stress-strain state must be taken into consideration in studying how the consolidation process proceeds in time and how the strength and stability of the soil mass are affected during the first stage.

It may be further pointed out that the aforesaid soil parameters can also be determined analytically, provided the parameters that describe the compressibility of the soil skeleton and pore water are known (see Chapter 2).

The *intermediate stage* is distinguished by intense changes in the stress-strain state of the soil mass in space and in time. This stage ends with almost total dissipation of excess pore pressure produced at the end of the initial stage, with the skeleton deformation entering the final stage of 'aged' creep that progressively slows down with time. The intermediate stage may be described by the following rheological parameters of the soil skeleton and pore water: compressibility coefficients m_{v1} and m_{v2}, creep parameter δ_{com}, moduli of deformation of skeleton K_m, K_l, G_m and G_l, parameters of skeleton creep δ_v and δ_γ, coefficient of compressibility of pore water m_w or modulus of compressibility of pore water $K_w = 3/m_w$, coefficient of permeability of pore water k or consolidation coefficient c_v and, finally, the non-dimensional parameter $\mu_c = \delta_{com}(H^2/c_v)$ which determines to what extent seepage and creep affect the consolidation process in the soil mass.

The *final stage* is distinguished by stabilisation of the transfer of external load to the soil skeleton, i.e., stabilisation of the stress-strain state. However, during this stage deformation may continue due to creep of the soil skeleton. As regards the shearing deformation, it is indirectly related to the dissipation of pore pressure and develops with time during consolidation as well as after it. In the final stage, deformation due to compression of skeleton proceeds at such a slow pace that it invites no resistance from pore water and, therefore, the deformation develops at an almost negligible hydraulic gradient. This phenomenon is observed in laboratory consolidation on clayey soils.

Hence, the final stage is described only by the rheological parameters of the soil skeleton, such as creep of the skeleton under one-dimensional consolidation m_{v3} or the moduli of deformation of the skeleton $K(t)$ and $G(t)$ corresponding to the 'aged' creep stage.

It is possible to determine the rehological parameters of soil required for describing the initial stage of development of the stress-strain state of the soil mass by consolidation tests on triaxial and constrained compression set-ups. The required tests may also be conducted on closed system torsion and shearing set-ups with or without the measurement of pore pressure. If pore pressure is not measured during the tests, then only the reduced soil parameters that describe its overall behaviour can be determined.

The elastic and overall moduli of deformation can be determined on the basis of the initial and final (conditionally stabilised) values of strain:

$$G_{red}^e = \frac{\tau}{\gamma^e}; \quad G_{red} = \frac{\tau}{\gamma}; \quad K_{red}^e = \frac{\sigma}{3\varepsilon^e}; \quad K_{red} = \frac{\sigma}{3\varepsilon}, \qquad \dots (5.2)$$

The moduli of plastic deformation can now be determined:

$$\frac{1}{G_{red}^p} = \frac{1}{G_{red}} - \frac{1}{G_{red}^e}; \quad \frac{1}{K_{red}^p} = \frac{1}{K_{red}} - \frac{1}{K_{red}^e}. \qquad \dots (5.3)$$

The overall creep parameters for the soil δ_γ' and δ_v' can be determined from the following expressions by comparing them with the experimental curves.

$$\left. \begin{array}{c} \gamma(t) = \tau \left\{ \dfrac{1}{G_{red}^e} + \dfrac{1}{G_{red}^p}[1 - \exp(-\delta_\gamma')t] \right\} \\[4mm] \varepsilon(t) = \dfrac{\sigma}{3} \left\{ \dfrac{1}{K_{red}^e} + \dfrac{1}{K_{red}^e}[1 - \exp(-\delta_v')t] \right\} \end{array} \right\} \qquad \dots (5.4)$$

and

It is evident from the above that δ_γ' and δ_v' differ from δ_γ and δ_v which describe the creep of skeleton.

The soil parameters under constrained compression are determined in a similar manner:

$$m_v = \varepsilon_1/\sigma_1; \quad m_{v1} = \varepsilon_1^e/\sigma_1; \quad m_{v2} = m_v - m_{v1}. \qquad \dots (5.5)$$

The creep parameter under constrained compression may be determined from the following expression:

$$\varepsilon_1(t) = \sigma_1 \left\{ m_{v1} + m_{v2}[1 - \exp(-\delta'_{com})t] \right\} \qquad \ldots (5.6)$$

This is done through comparison with the experimental creep curve, i.e.,

$$\delta'_{com} = \frac{1}{t} \ln \left[\frac{\varepsilon_1(t)}{\sigma m_{v2}} - \frac{m_{v1}}{m_{v2}} - 1 \right]. \qquad \ldots (5.7)$$

Should it be necessary to determine pore pressure and effective stresses in the skeleton in the initial stage of development of the stress-strain state, the parameters pertaining to the skeleton and pore water ought to be determined separately. This is not possible without first solving the problem of redistribution of total stresses between the skeleton and pore water, particularly while determining δ_γ, δ_v and δ_{com}. As regards the parameters of skeleton deformability, they can be determined easily, provided the pore pressure values at the commencement and completion of the test are known:

$$K^e = \frac{\sigma - u_w(0)}{3\varepsilon(0)}; \quad K = \frac{\sigma - u_w(\infty)}{3\varepsilon(\infty)};$$

$$m_{v1} = \varepsilon_1(0)/[\sigma_1 - u_w(0)]; \quad m_v = \varepsilon_1(\infty)/[\sigma_1 - u_w(\infty)];$$

$$\frac{1}{K^p} = \frac{1}{K} - \frac{1}{n^e}; \quad m_{v2} = m_v - m_{v1};$$

$$m_w = \varepsilon_1(0)/[u_w(0)n] = \varepsilon_1(\infty)/[u_w(\infty)n];$$

$$K_w = nu_w(0)/[3\varepsilon(0)] = nu_w(\infty)/[3\varepsilon(\infty)].$$

To determine the parameters of skeleton creep, δ_{com} and δ_v, it is necessary to utilise the solutions to the problem of redistribution of total stresses between soil skeleton and pore water, taking into consideration the equation of state adopted. For example, if the equation of state of the skeleton can be represented by the equation of hereditary creep (2.28) and pore water can be assumed to be linearly compressible, then the solution to the problem leads to the following equations:

—for constrained compression

$$u_w(t) = u_w(0) \left[1 + \frac{m_{v2}}{m_{v1}(1 + \chi r) + \chi m_{v2}} \right.$$

$$\left. \times \left\{ 1 - \exp \left[-\delta_{com} \left(1 - \frac{m_{v2}}{m_{v1} + nm_w} \right) t \right] \right\} \right], \quad \ldots (5.8)$$

where

$$u_w(0) = \frac{\sigma m_{v1}}{nm_w + m_{v1}}; \quad \chi = \frac{1}{e(0)}; \quad r = \frac{m_{v1}}{m_w}[1 + e(0)];$$

—for triaxial compression

$$u_w(t) = u_w(0)\left[1 + \frac{1/K^p}{(1+K^e)(1+\chi_1 r_1) + \chi_1/K^p}\right.$$

$$\left. \times \left\{1 - \exp\left[-\delta_v\left(1 - \frac{1/K^p}{1/K^e + n/K_w}\right)t\right]\right\}\right], \quad \dots \ (5.9)$$

where

$$r_1 = \frac{K_w}{K^e}[1 + e(0)]; \quad u_w(0) = \sigma\frac{K_w}{nK^e + K_w};$$

$$\chi_1 = \frac{3}{e(0)[1 + 2\xi]}.$$

It is evident from eqns. (5.8) and (5.9) that in a closed system the rate of growth of pore pressure is related not only to creep parameters δ_{com} and δ_v, but also to the deformation characteristics of soil skeleton and pore water. Parameters δ_{com} and δ_v can be found from a comparison of the analytical and experimental data.

To describe the intermediate state of the stress-strain state of the soil mass, the rheological properties of saturated soil may be determined from the results of tests conducted on a constrained compression set-up of the open system type, i.e., with provision for drainage. It is also possible to determine these parameters on a triaxial set-up but this is beset with procedural difficulties and the processing of test results is also more complicated. The reason for these difficulties lies in the variation of effective stresses along the height of the specimen during consolidation; this produces a non-homogeneous strain field which is almost impossible to control.

At present, there are no methods for determining consolidation parameters under compound stressed state. An approximate solution may be obtained by testing spherical specimens with central or surface drainage under all-round isotropic compression. Good results may also be achieved by testing cylindrical specimens under compression and torsion with one-way drainage. Finally, in laboratory conditions it is possible to conduct large-scale tests in a drum under constrained compression with axisymmetrical and three-dimensional drainage. Obviously, in this case it would be impossible to describe the stress-strain state of soil mass on the basis of reduced parameters.

Let us consider the case of constrained compression. According to the theory of hereditary creep, the rheological parameters of saturated soil are m_{v1}, m_{v2}, m_{v3}, c_v, δ_{com}, m_{wp}, n_c and μ_c.

A specimen of height h with initial coefficient of porosity $e(0)$ and degree of saturation $S_r(0)$ is subjected to a consolidating load $\sigma = q$. Free seepage is observed on the top surface of the specimen. However, there is no seepage at the bottom face. The total strain $\varepsilon_1(t) = S(t)/h$ and the pore pressure at the bottom, non-draining face are monitored during the course of consolidation.

The results of such tests have been represented in the form of $u_w(t)$ vs $\ln(t)$ and $\varepsilon_1(t)$ vs $\ln(t)$ curves (see Fig. 4.10).

The coefficient of relative compressibility and the coefficient of pore pressure are determined from the initial strain by the following expressions:

$$m_{v1} = \varepsilon_1(0)/[\sigma - u_w(0)]; \quad \beta_w(0) = u_w(0)/\sigma. \qquad \dots (5.10)$$

The sum of coefficients $m_{v1} + m_{v2} = \varepsilon_1(t_f)/\sigma$ can be found from the intermediate value of strain when the dissipation of pore pressure is over. In view of the above, it follows that

$$m_{v2} = \varepsilon_1(t_f)/\sigma - m_{v1}. \qquad \dots (5.11)$$

The creep parameter δ_{com} is determined from the solution to the one-dimensional problem of consolidation, taking into account creep of the skeleton and compressibility of the pore water [17]. This is possible if it is assumed that at the time the pore pressure attains its extreme value, the exponent of the exponential function is approximately equal to 10, i.e., $N_1 t_{max} = 10$, where

$$\left. \begin{aligned}
N_1 &= Q_1 + \sqrt{Q_1^2 - R_1^2}; \\
Q_1 &= \frac{1}{2}\left(\delta_{com} \frac{m_{v1} + nm_w + m_{v2}}{m_{v1} + nm_v} + \frac{c_v \pi^2}{h^2} \right); \\
R_1 &= \delta_{com} \frac{c_v \pi^2}{h^2}; \quad c_v = \frac{k}{\gamma_w(m_{v1} + nm_w)}.
\end{aligned} \right\} \qquad \dots (5.12)$$

It is evident from the above that a transcendental equation that includes known parameters m_{v1}, m_{v2}, m_w and k is obtained with respect to δ_{com}. The coefficient of permeability can be determined independently from the results of permeability tests and indirectly through the coefficient of consolidation. For example, if it is assumed that at time $t = t_f$, the exponent of the exponential function in the solution of the one-dimensional problem of consolidation [17] is equal to 10, i.e., $M_1 t_{max} = 10$, then it is possible to first determine c_v and then k. Moreover,

$$M_1 = Q_1 - \sqrt{Q_1^2 - R_1^2}. \qquad \dots (5.13)$$

However, according to eqns. (5.12), it is necessary in this case to assume that δ_{com} is known. Consequently,

$$k = \gamma_w(m_{v1} + nm_w)c_v. \qquad \dots (5.14)$$

It is noteworthy that the author has represented the coefficient of consolidation in a manner different from the commonly adopted form $c_v = k/\gamma_w m_v$, because it was necessary to take into account the creep of soil skeleton and compressibility of pore water. Moreover, the concept of coefficient of consolidation is somewhat wider in the given case since the mathematical expression for the degree of consolidation differs significantly from the classical solution of the Terzaghi-Gersevanov theory of consolidation. The mathematical expressions in the given case are as follows:

—for pore pressure

$$u_w(t,z) = \frac{4\sigma}{\pi} \sum_{n=1,3,5}^{\infty} \frac{1}{n}(C_n e^{-M_n t} - D_n e^{-N_n t}) \times \sin\frac{\pi n z}{h}; \qquad \dots \text{(5.15)}$$

—for degree of consolidation

$$\left.\begin{array}{l} U_{\mathrm{I}}(t) = 1 - \dfrac{8}{\pi^2} \sum_{n=1,3,\dots}^{\infty} \dfrac{1}{n^2}\left(C_n e^{-M_n t} - D_n e^{-N_n t}\right); \\[2mm] U_{\mathrm{II}}(t) = 1 - e^{-\delta_{\mathrm{com}} t} - \dfrac{8}{\pi^2} \sum_{n=1,3,\dots}^{\infty} \dfrac{1}{n^2}\left[\dfrac{C_n}{-M_n + \delta_{\mathrm{com}}}\right. \\[2mm] \left. \times \left(e^{-M_n t} - e^{-\delta_{\mathrm{com}} t}\right) - \dfrac{D_n}{N_n + \delta_{\mathrm{com}}}\left(e^{-N_n t} - e^{-\delta_{\mathrm{com}} t}\right)\right], \end{array}\right\}$$

$$\dots \text{(5.16)}$$

where

$$C_n = \frac{\delta_{\mathrm{com}} m_{v2} - [(m_{v1} + n m_w)N'_n + \delta_{\mathrm{com}} m_{v2} + k\alpha_n^2/\gamma_w]}{2(m_{v1} + n m_w)\sqrt{Q_n^2 + R_n^2}};$$

$$D_n = \frac{\delta_{\mathrm{com}} m_{v2} - [(m_{v1} + n m_w)M'_n + \delta_{\mathrm{com}} m_{v2} + k\alpha_n^2/\gamma_w]}{2(m_{v1} + n m_w)\sqrt{Q_n^2 + R_n^2}};$$

$$Q_n = \frac{1}{n}\left[\delta_{\mathrm{com}}\frac{m_{v1} + m_{v2} + n m_w}{m_{v1} + n m_w} + c_v\alpha_n^2\right];$$

$$R_n = \delta_{\mathrm{com}} c_v \alpha_n^2; \quad \alpha_n = n\pi/h; \quad M'_n = -Q_n + \sqrt{Q_n^2 - R_n^2};$$

$$N'_n = -Q_n - \sqrt{Q_n^2 - R_n^2}; \quad M_m = -M'_m; \quad N_m = -N'_m.$$

Thus, the mathematical description of the process of consolidation is rendered much more complicated if the creep of skeleton and compressibility of pore water are taken into account. This also introduces significant changes in the conventional parameters and coefficients, which necessitates the use of special testing methods and techniques for processing experimental results in order to determine the soil parameters during consolidation.

In this connection, attention is drawn to the non-dimensional coefficient $R_n = \delta_{\mathrm{com}} c_v \alpha_n^2$ which is similar to coefficient μ_c discussed in Chapter 1. Coefficient R_n describes the integral viscosity of the soil mass, i.e., it determines the duration of the consolidation process, subject to the assumptions that the soil skeleton experiences creep, the pore water is compressible and the coefficient of permeability is constant.

The problem of describing the consolidation process and determining the rheological parameters of soil is further complicated if the non-linear behaviour of soil skeleton and pore water is taken into consideration. Therefore, prior to undertaking the determination of the rheological parameters of soil, it is essential to assess whether it is necessary to consider the rheological properties in describing the consolidation process. Parameter μ_c can serve as the criterion for assessing whether the creep characteristics should be considered while dealing with the consolidation process. The consolidation index proposed by N.N. Maslov may also serve as a criterion:

$$t_1/t_2 = (h_1/h_2)^{n_c}. \qquad \qquad \dots (5.17)$$

For determination of n_c two tests are conducted on identical specimens of different heights, h_1 and h_2, and the times when both specimens achieve the same degree of consolidation are recorded. The following formula is then employed for determining n_c:

$$n_c = \frac{\ln(t_2/t_1)}{\ln(h_2/h_1)}. \qquad \qquad \dots (5.18)$$

If it is found that $n_c < 2$ during consolidation due to drainage, then it is necessary to take into account the effect of creep; if $n_c \approx 2$, then this is not necessary. Consolidation index n_c depends on soil consistency, degree of saturation and percentage of clayey particles in soil. According to the data reported by N.N. Maslov [13], the index of consolidation depends on plasticity index I_p and liquidity index I_L, as shown in Fig. 5.1.

In conclusion, the author would like to discuss another important question concerning predication of the settlement of foundations in two-dimensional problems, when the consolidation and shearing deformations grow side by side.

If the shearing deformation depends on the degree of consolidation of soil, i.e., on the soil density and effective normal stress, then it will be governed by the rate of the consolidation process. Otherwise, shearing deformation will develop independently because shearing stresses remain almost constant during consolidation. In view of this, it suffices to determine the rheological parameters of the soil skeleton in shear in accordance with the chosen rheological model of soil.

Let us examine the case wherein the rheological parameters of soil are a function of the density of soil skeleton and the effective normal stresses. For the relation $\dot{\gamma} = \tau/[\eta^v(\sigma', \varphi)]$, the rate of shearing deformation differs at individual points of the soil mass subjected to consolidation. It partially tends to level out towards the end of the consolidation process but that, too, not fully because σ and φ depend on the co-ordinates. In the final stage, the consolidation process proceeds at a constant rate of settlement which is determined by the shearing deformation due to viscous flow. If the relation between shearing creep and σ'

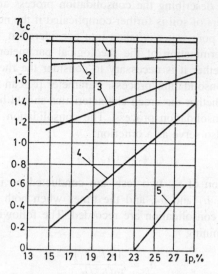

Fig. 5.1. Variation of n_c with I_P and I_L for clayey soil (data from M.N. Maslov).

$1—I_L = 0.88$; $2—I_L = 0.62$; $3 - I_L = 0.37$; $4—I_L = 0.12$; $5—I_L = 0.08$.

and φ is more complex, as given below,

$$\dot{\gamma} = \frac{\tau - \sigma' \cdot \operatorname{tg} \varphi - c(\varphi)}{\eta(\sigma, \varphi)}, \qquad \dots (5.19)$$

then the rate of shearing deformation at various points of the soil mass under consolidation can vary in a wide range. It can also have a zero value because σ' and φ increase with time and the numerator of the fraction may become zero. Obviously, in this case it becomes necessary to determine the relations $c(\varphi)$ and $\eta(\sigma, \varphi)$, which is possible only on the basis of the results of consolidated-undrained tests conducted at various consolidating loads.

It follows from the foregoing that the settlement of the foundation of a structure under a local load will develop with time in conformity with the process of consolidation as well as the shearing creep of the foundation soil. It may be pointed out that although shearing creep develops under the influence of shearing stresses, it depends on consolidation.

The rheological parameters of soil required for evaluating the stress-strain state at the final stage of consolidation can be determined on the basis of the results of consolidated-drained tests conducted on triaxial compression, constrained compression, torsion and shearing set-ups.

The final stage is distinguished by absence of pore pressure in the soil (when the creep process enters the final stage) and may culminate either in total damping of strain or in its development in direct proportion to the logarithm of time. An even greater complexity of the creep process is possible under the action of local load.

In full-scale tests, the settlement of the foundation continues even after outflow of water has ceased. This is due not as much to the volumetric creep of soil, as to the shearing creep. Under constrained compression, this process can be distinctly observed after total dissipation of pore pressure and is known as secondary consolidation. It may be described by a single parameter $m_{v3} = a \ln(t/t_f)$ (where a is the tangent of the angle of inclination of the $\varepsilon(t)$ vs $\ln(t)$ line in the range $t > t_f$). Thus, the final stage of one-dimensional consolidation can be evaluated without difficulty.

In the case of two- or three-dimensional consolidation under local load, the rheological parameters of soil should be determined both for volumetric strain and distortion. In the final stage of consolidation, the volumetric deformation ceases to play a governing role and the settlement develops mostly due to shearing creep of the skeleton. Consequently, it becomes necessary to determine the parameters of long-term deformation under consolidation as well as shearing. As already discussed, the parameters that describe long-term deformation under consolidation are m_{v3} and a. It is more difficult to determine the parameters of shearing creep because the latter depends on the rheological model selected.

A number of models are possible, including viscous flow model $\dot{\gamma} = \tau/\eta^v(t)$, viscoplastic flow model $\dot{\gamma} = (\tau - \tau^*)/\eta^{vp}(t)$ and so forth. In a particular case, when $\eta^v(t) = \eta_0^v(1 + \alpha t)$, the shearing deformation develops with time in accordance with the logarithmic rule, i.e.,

$$\gamma(t) = \frac{\tau}{\eta_0 \alpha} \ln(1 + \alpha t), \qquad \dots (5.20)$$

where $t > t_f$.

Parameters η_0 and α may be determined from the creep curve plotted on the basis of the results of consolidated-undrained tests under constant applied shearing stress.

6

Creep of Partially Saturated Clayey Soils

1. General Principles

A soil mass containing partially saturated layers which, upon interaction with the foundation of the structure undergo deformation in accordance with the laws of rheology of single-phase media, is known as a partially saturated base.

The rheological properties of partially saturated soils may manifest themselves in different ways in the foundations of structures, depending on the structure and composition of the soil mass, isotropicity and homogeneity of the soil, initial stress-strain state, dimensions of the foundation and loading conditions. Therefore, in each particular case, it is necessary to justify the selected geomechanical model of the foundation for evaluation of the rheological processes occurring in space and time due to interaction between the soil and the foundation of the structure.

2. Analysis of Time-dependent Settlement of Partially Saturated Bases

Rheological processes manifest themselves in foundation soils in different ways, depending on the nature of development of the stress-strain state in the soil mass, which is interacting with the structure. The consolidation strain plays the governing role in some cases and the shearing strain in others. For example, consolidation strain predominates in the soil under foundations of width 10 m or more subjected to loads of up to 0.3 MPa. On the other hand, shearing strain predominates in soil under foundations of width 3 m or less subjected to a load of 0.3 MPa. When the width of foundation lies between 3 and 10 m, the consolidation and shearing strains are almost equal. According to N.A. Tsytovich [22], the stress-strain state of the foundations of structures can be divided into two phases: volumetric strain predominates in the first and shearing strain in the second. Obviously, such a division is somewhat contrived because under the action of local loads both volumetric and shearing strains develop simultaneously in the soil mass from the very beginning.

Let the total settlement of the surface of the soil base be represented as a sum of linear strains consisting of volumetric and shearing strains along an arbitrary vertical,

$$s = s_v + s_\gamma = \int_0^z \varepsilon_z dz = \int_0^z \varepsilon(z)dz + \int_0^z e(z)dz, \qquad \ldots \text{(6.1)}$$

where

$$\left.\begin{array}{c} \varepsilon_z = \varepsilon(z) + e(z) = \dfrac{\sigma(1 - 2\mu_0)}{E_0} + \dfrac{\sigma_z - \sigma}{2G_0}; \\[2mm] \varepsilon = (\varepsilon_x + \varepsilon_y + \varepsilon_z)/3; \quad \sigma = (\sigma_x + \sigma_y + \sigma_z)/3. \end{array}\right\} \qquad \ldots \text{(6.2)}$$

The expression of the rate of settlement can be written in a similar manner as follows:

$$\dot{s} = \dot{s}_v + \dot{s}_\gamma = \int_0^z \dot{\varepsilon}_z dz = \int_0^z \dot{\varepsilon}(z)dz + \int_0^z \dot{e}(z)dz, \qquad \ldots \text{(6.3)}$$

where

$$\left.\begin{array}{c} \dot{\varepsilon}_z = \dot{\varepsilon}(z) + \dot{e}(z) = \dot{\chi}_v \sigma + (\sigma_z - \sigma)\dot{\chi}_\gamma; \\[2mm] \dot{\chi}_v = \dot{\varepsilon}/\sigma = f_v(\sigma, \tau_i, K_\sigma, t)\sigma; \\[2mm] \dot{\chi}_\gamma = \dot{\gamma}_i/(2\tau_i) = f_\gamma(\tau_i, \sigma. K_\sigma, t)/(2\tau_i). \end{array}\right\} \qquad \ldots \text{(6.4)}$$

Because of the complex spatial-temporal nature of the $\sigma_z(x, y, z, t)$ and $\sigma(x, y, z, t)$ distributions, the development of settlements of the surface of the soil mass is also complicated. Moreover, the σ_z and σ distributions must be obtained on the basis of the solution of the given problem in accordance with the rheological model of the soil base. However, a closed form solution to this problem is possible only for the particular case when the base is represented by homogeneous soils with isotropic properties and linear stress-strain relationship.

This complex problem can be resolved in the following manner. The stress distribution corresponding to the solution for homogeneous, linearly deformable half-space is adopted as the first approximation. As regards deformations, they can be determined on the basis of the known state of stress by applying the rheological equations of state (6.2) or (6.4).

The above method of evaluating the stress-strain state of the base is an approximate method but conforms with that presently adopted in SNIP. The distinguishing feature of the proposed method is that it takes into account the volumetric and shearing strains as well as the rheological properties of the soil base while determining the total settlement, also in conformity with the requirements laid down in SNIP 2.02.01-83. Obviously, this does not exclude the use of numerical methods in necessary cases for the determination of the stress-strain state since it then becomes possible to consider numerous factors that affect the rheological process.

The established solutions for the theory of elasticity (at $x = y = 0$) [20] can be applied for analysis and evaluation of the stress-strain state of the soil base under the action of a local load:
—for a two-dimensional problem the stresses along the z-axis are:

$$\left.\begin{array}{l} \sigma_z(z) = \dfrac{2q}{\pi}\left(\operatorname{arctg}\dfrac{b}{z} + \dfrac{bz}{b^2 + z^2}\right); \\[3mm] \sigma(z) = \dfrac{2q}{3\pi}(1 + \mu_0)\operatorname{arctg}\dfrac{b}{z}, \end{array}\right\} \qquad \ldots (6.5)$$

where b is the half-width of the loaded strip and q is the intensity of load.
—for an axisymmetrical problem the stresses along the z-axis are:

$$\left.\begin{array}{l} \sigma_z(z) = q\left[1 - \dfrac{z^3}{(b^2 + z^2)^{3/2}}\right]; \\[3mm] \sigma(z) = \dfrac{2}{3}q\left[1 - \mu_0 - (1 - \mu_0)\dfrac{z}{(b^2 + z^2)^{1/2}}\right], \end{array}\right\} \qquad \ldots (6.6)$$

where b is radius of the loaded area.
—for a rectangular loaded area the stresses along the z-axis are:

$$\left.\begin{array}{l} \sigma_z = \dfrac{q}{2\pi}\left[\operatorname{arctg}\dfrac{m}{m\sqrt{1 + m^2 + n^2}} + \dfrac{mn(1 + m^2 + n^2)}{(m^2 + n^2)(1 + m^2)\sqrt{1 + m^2 + n^2}}\right]; \\[3mm] \sigma = \dfrac{q(1 + \mu_0)}{3\pi}\operatorname{arctg}\dfrac{n}{m\sqrt{1 + m^2 + n^2}} \end{array}\right\}$$

$$\ldots (6.7)$$

where $n = b/l$, $m = z/(2b)$, b and l represent the width and length of the rectangle respectively.

To facilitate the calculations, the values of σ_z/q and $3\sigma/q$ have been reported in [20] along various verticals of the area under load for both two- and three-dimensional problems. On the basis of formulae (6.5) and (6.6), it is possible to analyse the stress-strain state of the soil base and estimate the relative proportions of the shearing and volumetric strains at different depths from the surface.

$$K_{\gamma v} = l_z(z)/\varepsilon(z) = C(\sigma_z - \sigma)/\sigma, \qquad \ldots (6.8)$$

where $C = (1 + \mu_0)/[(1 - 2\mu_0)3]$.

Coefficient $K_{\gamma v}$ depends on the deviation of the deformed state of soil at the given point from the conditions of all-round isotropic compression. For example, for a plane strain problem, in accordance with eqn. (6.5), we have

$$K_{\gamma v} = \frac{e_z}{\varepsilon} = C\frac{\sigma_z - \sigma}{\sigma} = C\left(\frac{\sigma_z}{\sigma} - 1\right)$$

$$= \frac{3}{1 + \mu_0} \times \left[1 + \frac{bz}{(b^2 + z^2)\operatorname{arctg}(b/z)}\right]C. \qquad \ldots (6.9)$$

The ratio $\varepsilon_z/\varepsilon_{z,com}$ can be introduced in a similar manner to describe how much the deformed state at the given point of the soil base deviates from the conditions of constrained compression. The above ratio corresponds to the empirical coefficient K_{δ_e} that is recommended in SNIP 2.02.01-83 to determine the subsidence of the soil base when the width of the foundation is less than 12 m. Parameters ε_z and $\varepsilon_{z,com}$ are correlated by the following expression:

$$\varepsilon_z\varepsilon_{z,com} = (1 + \mu_0 - 3\mu_0\sigma/\sigma_z)/\beta, \qquad \dots (6.10)$$

where $\beta = 1 - 2\mu_0^2/(1 - \mu)$.

As is evident from expression (6.9), the relation between $K_{\gamma v}$ and b is non-linear. At contact b, coefficient $K_{\gamma v}$ increases with depth (Fig. 6.1).

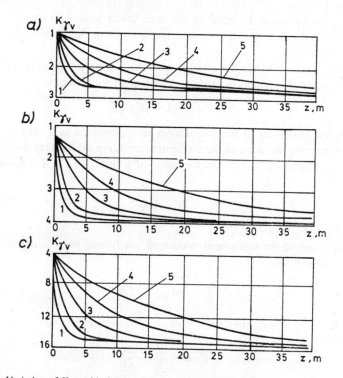

Fig. 6.1. Variation of $K_{\gamma v}$ with depth z for the foundation of a structure.

a—$\mu = 0.2$; b—$\mu = 0.3$; c—$\mu = 0.45$; 1—$b = 1$ m; 2—$b = 2$ m; 3—$b = 3$ m; 4—$b = 10$ m; 5—$b = 20$ m.

Thus, the contribution of shearing strain of soil base to the total settlement varies depending on the width of the loaded area. The wider the loaded area, the lesser the contribution of the shearing strain and vice versa.

Variation in the magnitude of constrained compression with the loaded area can be estimated in an exactly similar manner. By substituting parameters σ and σ_z from (6.5) in (6.10), it is found that

$$K_{com}(b, z) = \frac{1 - \mu_0^2}{1 - \mu_0 - 2\mu_0^2} \frac{bz/(b^2 + z^2) + (1 - \mu_0)\operatorname{arctg}(b/z)}{bz/(b^2 + z^2) + \operatorname{arctg}(b/z)}. \qquad \dots (6.11)$$

It is evident from the above equation that K_{com} depends on the width of the loaded area and the depth at which the layer is located. Consequently, constrained compression conditions are observed to prevail in foundations of structures in extremely rare cases. The stress-strain state of the foundation can be analysed for other types of loaded areas by means of eqns. (6.6) and (6.7).

The deviation is observed to be still greater for other lines, say, for example the line passing along the edge of the area under load. In view of the above, there is a need to improve the method for analysis of foundations that has been adopted in SNIP.

It is convenient to use the generalised formulae proposed by N.A. Tsytovich [22] for practical calculation of the settlement of foundations:

$$s = \omega q b (1 - \mu_0^2)/E_0. \qquad \dots (6.12)$$

Let the settlement of the foundation be represented as a sum $s = s_\gamma + s_v$ after determining the share of the settlement resulting exclusively due to shearing strain, by assuming $\mu = 0.5$. On the basis of formula (6.12) it is found that:

$$s_\gamma = \frac{\omega q b}{4G_0}, \qquad \dots (6.13)$$

where $G_0 = E_0/[2(1 + \mu_0)]$.

The share of the settlement produced solely due to volumetric strain can now be determined:

$$s_v = \frac{\omega q b}{6k_0}(1 + \mu_0), \qquad \dots (6.14)$$

where $K_0 = E_0/[3(1 - 2\mu_0)]$.

Hence,

$$s = s_\gamma + s_v = \frac{\omega q b}{4G_0} + \frac{\omega q b}{6K_0}(1 + \mu_0). \qquad \dots (6.15)$$

It is not difficult to notice that the original equation (6.12) can be derived from formula (6.15). The ratios s_γ/s and s_γ/s_v can also be easily established on the basis of expressions (6.12) to (6.14):

$$\left. \begin{array}{l} s_\gamma/s = 1/[2(1 - \mu_0)]; \\ s_\gamma/s_v = 1/(1 - 2\mu_0). \end{array} \right\} \qquad \dots (6.16)$$

It follows from the above that $s_\gamma/s \to 1$ at $\mu_0 \to 0.5$ and $s_\gamma/s_v \to \infty$ because $s_v \to 0$. Different results and analytical formulae can be obtained depending on the rheological model of the soil base selected.

According to the theory of hereditary creep, the settlement of homogeneous soil base can be calculated fairly simply if it is assumed that the state of stress in the base does not change with time. This is possible if the creep kernels for volumetric strain $K_v(t, \tau)$ and distortion $K_\gamma(t, \tau)$ are similar. However, such cases are extremely rare. The same principle can be adopted with a fair degree of accuracy for practical calculations involving relatively more complex $K_v(t, \tau)$ and $K_\gamma(t, \tau)$ expressions. This is achieved by replacing G_0 and K_0 in formulae (6.15) with $G_0(t)$, $K_0(t)$ respectively:

$$\left.\begin{array}{l} \dfrac{1}{G_0(t)} = \dfrac{1}{G_m} + \dfrac{1}{G_l}(1 - e^{-\delta_\gamma t}); \\[2mm] \dfrac{1}{K_0(t)} = \dfrac{1}{K_m} + \dfrac{1}{K_l}(1 - e^{-\delta_v t}). \end{array}\right\} \qquad \dots (6.17)$$

When $K_0(t) = K_m = $ const., it is possible to represent $G_0(t)$ through the following relation:

$$\frac{1}{G_0(t)} = \frac{1}{G_m} + \frac{1}{G_l}\ln(t/t_1), \qquad \dots (6.18)$$

where t_1 is a parameter with units of time.

In the case of ideal viscosity under shear and elasticity during volumetric strain, the rate of settlement of a circular loaded plate is given by the expression [1.17]:

$$\dot{s} = \frac{\pi q r_0}{2\eta^v}\left\{4 - 3\exp[-3K_0 t/(4\eta^v)]\right\}, \qquad \dots (6.19)$$

where r_0 is the radius of the foundation and η^v is the viscosity under shear.

Obviously, at $t \to \infty$, formula (6.19) leads to the following expression:

$$\dot{s} = 2\pi q r_0/\eta^v, \qquad \dots (6.20)$$

whereby at $q = 0.3$ MPa, $r_0 = 10$ m and $\eta^v = 10^{10}$ MPa. sec, it is found that $\dot{s} = 2 \times 3.14 \times 0.3 \times 1000/10^{10} = 1.884 \times 10^{-7}$ cm/s, which leads to a settlement of 0.585 cm per year and 5.85 cm in 10 years.

Expression (6.20) can be applied not only for predicting the rate of settlement of circular foundations, but also for determining the viscosity of soil on the basis of plate load tests, which must be conducted until the state $\dot{s} = $ const. is reached. From these tests, the viscosity can be determined by the relation $\eta^v = 2\pi q r_0/\dot{s}$.

Following the theory of viscoplastic flow, the settlement of a homogeneous base can be found by the method of layer-wise summation of strain, taking into account the stress state determined in accordance with the theory of a linearly elastic medium. This is done by means of eqns. (6.3) and (6.4), wherein integration is replaced by summation:

$$\dot{s} = \sum_{j=1}^{j=n} \left\{ \sigma(z_j)\dot{\chi}_v(z_j) + [\sigma_z(z_j) - \sigma(z_j)]\dot{\chi}_\gamma(z_j) \right\} \Delta h_j, \qquad \ldots \ (6.21)$$

where $\sigma(z_j)$ and $\sigma_z(z_j)$ are the values of σ and σ_z respectively in the jth layer of thickness Δh_j; $\dot{\chi}_v(z_j)$ and $\dot{\chi}_y(z_j)$ are the values of functions $\dot{\chi}_v$ and $\dot{\chi}_\gamma$ respectively in the jth layer.

If follows from the above that the rate of settlement is different in each elementary layer and depends on the stresses in the given layer.

To determine the functions $\dot{\chi}_v$ and $\dot{\chi}_\gamma$ it is necessary to know $\sigma(z_j)$ and $\tau_i(z_j)$ or $\sigma_i(z_j)$ in each elementary layer. In a two-dimensional problem the shearing stresses under strip loading can be found from the formula:

$$\tau_i = \frac{\sigma_1 - \sigma_3}{\sqrt{3}} = \frac{2q}{\pi\sqrt{3}} \sin\alpha, \qquad \ldots \ (6.22)$$

where α is the angle of visibility.*

The normal and octahedral shearing stresses can be determined in a similar manner:

$$\left. \begin{aligned} \sigma_i &= \sigma_1 - \sigma_3 = \frac{2q}{\pi}\sin\alpha; \\ \tau_m \frac{\sqrt{2}}{3}(\sigma_1 - \sigma_3) &= \frac{2\sqrt{2}}{3\pi} q \sin\alpha. \end{aligned} \right\} \qquad \ldots \ (6.23)$$

Similar expressions can also be written for the strain invariants:

$$\left. \begin{aligned} \gamma_i &= \frac{2}{\sqrt{3}}(\varepsilon_1 - \varepsilon_3); \\ \varepsilon_i &= \frac{2}{3}(\varepsilon_1 - \varepsilon_3); \quad \gamma_0 = \frac{2\sqrt{2}}{3}(\varepsilon_1 - \varepsilon_3). \end{aligned} \right\} \qquad \ldots \ (6.24)$$

The soil may experience deformation due to shearing creep both when $\tau_i < \tau_i^*$ and when $\tau_i > \tau_i^*$. In the first case, the soil experiences viscous flow with a viscosity $\eta^v(\sigma)$, while in the second case, it experiences viscoplastic flow with the viscosity $\eta^{vp}(\sigma)$.

Suppose the soil skeleton displays viscous behaviour at $\tau_i < \tau_i^*$ and viscoplastic behaviour at $\tau_i > \tau_i^*$, i.e.,

$$\dot{\gamma}_i = \frac{\tau_i}{\eta^v(\sigma)}; \quad \dot{\gamma}_i = \frac{\tau_i - \tau_i^*}{\eta^{vp}(\sigma)}. \qquad \ldots \ (6.25)$$

---

* In other words, the angle of inclination to vertical of the radial line through the point of interest—Technical Editor.

Volumetric strain of the skeleton is described by equations of the type:

$$\begin{aligned}\left.\begin{aligned}\dot{\varepsilon} &= \frac{\sigma - \sigma^*}{\eta_v(t)} = \frac{\sigma - \sigma^*}{\alpha t} \text{ at } \sigma > \sigma^*;\\ \dot{\varepsilon} &= 0 \text{ at } \sigma < \sigma^*.\end{aligned}\right\}\end{aligned} \qquad \dots (6.26)$$

Substituting these expressions in formula (6.4), it is found that:

$$\left.\begin{aligned}\dot{\varepsilon}_z &= \frac{1 - \sigma^*/\sigma}{\alpha t} + \frac{\sigma_z - \sigma}{2\eta^v(\sigma)} \text{ at } \tau_i < \tau_i^*;\\ \dot{\varepsilon}_z &= \frac{1 - \sigma^*/\sigma}{\alpha t} + \frac{\sigma_z - \sigma}{2\eta^{vp}(\sigma)}(1 - \tau_i^*/\tau_i) \text{ at } \tau_i > \tau_i^*.\end{aligned}\right\} \qquad \dots (6.27)$$

Taking into account eqns. (6.27), the following relations are obtained from formula (6.21):

at $\tau_i < \tau_i^*$

$$\dot{s} = \sum_{j=1}^{j=n} \left\{ \frac{1 - \sigma^*/\sigma(z_j)}{\alpha t} + \frac{\sigma_z(z_j)}{2\eta^{vp}\sigma(z_j)}[1 - \tau_j^*/\tau_i(z_j)] \right\} \Delta h_i; \qquad \dots (6.28)$$

at $\tau_i \geq \tau_i^*$

$$\dot{s} = \sum_{j=1}^{j=n} \left\{ \frac{1 - \sigma^*/\sigma(z_j)}{\alpha t} + \frac{\sigma_z(z_j) - \sigma(z_j)}{2\eta^{vp}\sigma(z_j)}[1 - \tau_j^*/\tau_i(z_j)] \right\} \Delta h_i. \qquad \dots (6.29)$$

While deriving the above relations, it was kept in mind that:

$$\left.\begin{aligned}\eta^v(\sigma) &= \eta^v(1 + K^v\sigma);\\ \eta^{vp}(\sigma) &= \eta^{vp}(1 + K^{vp}\sigma).\end{aligned}\right\} \qquad \dots (6.30)$$

From the relations describing the rate of settlement, the expressions for settlement itself are obtained as follows:

at $\tau_i < \tau_i^*$

$$s(t) = \sum_{j=1}^{j=n} \left[\frac{1 - \sigma^*/\sigma(z_j)}{\alpha} \ln(t/t_0) + t\frac{\sigma_z(z_j) - \sigma(z_j)}{2\eta^v\sigma(z_j)} \right] \Delta h_i; \qquad \dots (6.31)$$

at $\tau_i \geq \tau_i^*$

$$\dot{s}(t) = \sum_{j=1}^{j=n} \left[\frac{1 - \sigma^*/\sigma(z_j)}{\alpha} \ln(t/t_0) + t\frac{\sigma_z(z_j) - \sigma(z_j)}{2\eta^{vp}\sigma(z_j)} \right] \Delta h_i. \qquad \dots (6.32)$$

Expressions (6.31) and (6.32) may be obtained in other forms when η^v and η^{vp} are time dependent. For example, if the time-dependent η^v and η^{vp} are described by the following relations:

$$\left.\begin{array}{c} \eta^v(\sigma,t) = \eta^v(1 + K^v\sigma)(1 + \beta^v t) \\ \eta^{vp}(\sigma,t) = \eta^{vp}(1 + K^{vp}\sigma)(1 + \beta^{vp}t), \end{array}\right\} \qquad \dots (6.33)$$

and

Then the following expressions for settlement are obtained from eqns. (6.28) and (6.29):

at $\tau_i < \tau_i^*$

$$s(t) = \sum_{j=1}^{j=n} \left\{ \frac{1 - \sigma^*/\sigma(z_j)}{\alpha} \ln(t/t_0) + \frac{\sigma_z(z_j) - \sigma(z_j)}{2\eta^v[1 + K^v\sigma(z_j)]\beta^v} \ln(1 + \beta^v t) \right\} \Delta h_j;$$

$$\dots (6.34)$$

at $\tau_i \geq \tau_i^*$

$$s(t) = \sum_{j=1}^{j=n} \left\{ \frac{1 - \sigma^*/\sigma(z_j)}{\alpha} \ln(t/t_0) \right.$$

$$\left. + \frac{\sigma_z(z_j) - \sigma(z_j)}{2\eta^{vp}[1 + K^{vp}\sigma(z_j)]\beta^{vp}} \ln(1 + \beta^{vp}t) \right\} \Delta h_j. \quad \dots (6.35)$$

Thus formulae have been derived for determining the settlement of a partially saturated soil base under the action of a local strip load, taking into account the rheological behaviour of soil under volumetric strain and distortion.

Before applying these formulae, it is necessary to determine the boundaries whereby the condition $\tau_i \geq \tau_i^*$ is satisfied. In a layer of soil base under compression there are two such boundaries separating the layer of viscoplastic flow from the layers of viscous flow between which former is sandwiched.

Let us illustrate with an example. Suppose a load of intensity $q = 0.5$ MPa is applied over a strip of width $2b = 10$ m on the surface of a homogeneous, isotropic half-space of soil. Suppose the soil is partially saturated and displays rheological behaviour under shear and volumetric deformation governed by formulae (6.25) and (6.26) whose parameters are known. The first step is to plot the σ_z, σ and τ_i diagrams along the axis of the loaded area with due consideration of the initial stress-strain state. Next, the boundaries of viscoplastic flow are determined by comparing τ_i with τ_i^*. In the example under consideration, the given values are: $c = 0.1$ MPa, $\varphi = 15°$, $\rho = 2$ g/cm^3 and the boundaries lie at the depths $z_1^* = 2$ m and $z_2^* = 14.5$ m. Finally, the total settlement of the layer can be determined by summation of the settlement with viscous and viscoplastic flow.

Let it be noted in conclusion that the above-mentioned method of layer-wise summation of strain due to creep of soil and its rheological behaviour may also be applied to non-homogeneous soil bases.

3. Effect of Hardening of Partially Saturated Soils on Initial Critical Load

According to N.A. Tsytovich [22], the stress-strain state of the foundation of a structure can be divided into two phases, according to the intensity of the applied loads. The first phase covers the emergence of slip zones in the soil base and completion of consolidation. During this phase conditions of limiting equilibrium appear in isolated regions at the edges of the loaded area. In the second phase, continuous regions of limiting equilibrium are formed under the loaded surface, the soil becomes unstable and the load-bearing capacity of the soil base is fully utilised. The load corresponding to the first phase is safe for the foundation and is known as the initial critical load. The second-phase load is known as the limiting critical load on the soil base.

At present, the determination of initial critical load is based on the solution to the two-dimensional problem of the theory of elasticity and the condition of limiting equilibrium without taking into consideration the hardening of soil due to consolidation of the base. The solution of the problem given below takes hardening into account and it shall be shown that when hardening is ignored, significant reduction in the initial critical load results.

Suppose a uniformly distributed load of intensity q is acting on a strip of width $2b$ on the surface of a homogeneous, partially saturated half space of soil. Let us determine the initial critical load q^*, which produces plastic deformation at a given depth under the edges of the loaded portion of soil whose properties are known.

In this case, the stress and strain components are determined on the basis of the known solutions [22] as follows:

$$\left.\begin{array}{l} \sigma_{1,2} = q(\alpha \pm \sin \alpha)/\pi; \\ \sigma = -2(1 + \mu_0)q(\alpha_1 - \alpha_2)/(3\pi); \\ \varepsilon = (\varepsilon_x + \varepsilon_y)/3 = \sigma/(3K_0), \end{array}\right\} \qquad \dots (6.36)$$

where

$$\alpha_1 = \text{arctg}\,\frac{y}{x-b}; \qquad \alpha_2 = \text{arctg}\,\frac{y}{x+b} \quad \text{at } x > b;$$

$$\alpha_1 = \frac{\pi}{2} + \text{arctg}\,\frac{y}{x-b}; \quad \alpha_2 = \text{arctg}\,\frac{y}{x+b} \text{ at } x < b.$$

The increment in density at an arbitrary point of the base can be determined on the basis of volumetric strain of soil by the formula:

$$\Delta\rho \approx \rho_0\varepsilon_v = \rho_0 \cdot 3\varepsilon. \qquad \dots (6.37)$$

Taking into account (6.36), it is found that

$$\Delta\rho = \rho_0 \cdot 2q(\alpha_1 - \alpha_2)(1 + \mu_0)/(3\pi K_0). \qquad \dots (6.38)$$

Thus a relation has been obtained for determining the density of soil at various points of the base. The results of calculations based on formula (6.38)

110

at $\rho_0 = 1.4$ g/cm^3, $K_0 = 5$ MPa and $\mu = 0.3$ are shown in Fig. 6.2. It is evident from these results that the stress-strain state of the base is non-homogeneous across the plane and has a central kernel under the loaded area. This factor affects the initial critical load but is not taken into account in the conventional analytical methods of its determination.

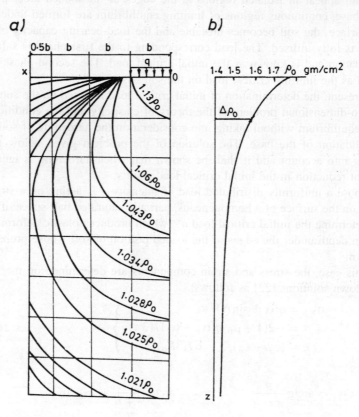

Fig. 6.2. (a) Density isolines in partially saturated base of foundation and (b) variation of density along depth z and $x = 0$.

Let us examine the stress-strain state of a partially saturated soil base at the final stage of the consolidation phase when a plastic flow zone appears under the edges of the segment under load. It will be assumed that at constant moisture content the hardening of soil during consolidation occurs due only to increase in cohesion; the angle of friction remains constant, i.e.,

$$\left.\begin{array}{l} \varphi(\rho) = \varphi = \text{const}; \\ c(\rho) = c_0(\rho_0) + c_\rho \Delta\rho. \end{array}\right\} \qquad \dots (6.39)$$

Given the above assumptions, the cohesion at various points of the consolidated mass is determined on the basis of expressions (6.38) and (6.39) as follows:

$$c(\rho) = c_0 + c_\rho \; \rho_0 \frac{2(1 + \mu_0)}{3\pi K_0} q(\alpha_1 - \alpha_2), \qquad \ldots (6.40)$$

or

$$c(\rho) = c_0 + K_\rho q(\alpha_1 - \alpha_2)/\pi.$$

For plane strain conditions, the equation of limiting equilibrium can be written on the basis of the Coulomb-Mohr theory in the following form:

$$\sigma_1 = \sigma_2 = [\sigma_1 + \sigma_2 + 2c(\rho)\,\mathrm{ctg}\,\varphi]\sin\varphi, \qquad \ldots (6.41)$$

where $c(\rho)$ is determined from formulae (6.40).

In this case, the critical load should be determined by simultaneously considering expressions (6.36) and (6.41) and taking (6.40) into account. Furthermore, it should be assumed that the working load and dead weight of soil produce a hydrostatic stress-strain state, i.e.,

$$\sigma_1 = \sigma_2 = \gamma(\varphi + h). \qquad \ldots (6.42)$$

As a result of the above operations, the following equations are obtained for the points where the condition of limiting equilibrium is satisfied:

$$y = \frac{q - \gamma h}{\pi \gamma} \left[\frac{\sin\alpha}{\sin\varphi} - \alpha(1 + K_\rho \cdot \mathrm{tg}\,\varphi) - \frac{c_0}{\gamma} \times \mathrm{ctg}\,\varphi - h \right]. \qquad \ldots (6.43)$$

When $k_\rho = 0$, eqn. (6.43) coincides with the well-known equation for determining the ordinates of the plastic flow zone without consideration of hardening. In determining y_{max}, eqn. (6.43) is differentiated with respect to α and the resultant expression is equated to zero:

$$y' = \frac{q\gamma h}{\pi \gamma} \left[\frac{\cos\alpha}{\sin\varphi} - (1 + K_\rho \cdot \mathrm{tg}\,\varphi) \right] = 0,$$

wherefrom, it follows that:

$$\alpha^* = \arccos[\sin\varphi(1 + K_\rho \cdot \mathrm{ctg}\,\varphi)]. \qquad \ldots (6.44)$$

Substituting the value of α^* found in eqn. (6.43), it is found that:

$$q^* = \pi \frac{\sin\varphi}{\sin\alpha^* - \alpha^* \cos\alpha^*} [\gamma(y_{max} + h) + c \cdot \mathrm{ctg}\,\varphi] + \gamma h. \qquad \ldots (6.45)$$

As a particular case, in the absence of hardening, i.e., when $K_\rho = 0$, the expression obtained earlier by N.P. Puzyrevskii for q^* is obtained because $\alpha^* = (\pi/2) - \varphi$.

Thus a solution has been obtained for determining the critical load on a soil base for a given maximum depth of plastic flow zone y_{max}.

In the case of heavy clays distinguished by $\varphi \approx 0$, expression (6.45) can be written as follows:

$$q^* = \frac{\pi c_0}{\sin \alpha^* - K_\rho \alpha^*} + \gamma h. \qquad \qquad \dots (6.46)$$

Since α^* depends upon hardening or the accumulated volumetric strain during the consolidation phase, it is obvious that q^* will depend on the degree of consolidation of the soil base. If the soil parameters are taken as $\rho_0 = 1.6$ g/cm^3, $\mu_0 = 0.3$, $K_0 = 10$ MPa, $c_0 = 0.1$ MPa and $c_\rho = 1$ MPa, it is found that $K_\rho = 0.139$, $\alpha^* = 1.4311$ rad. $\sin \alpha^* = 0.9903$ and $q^* = 0.4168$ MPa. If the soil hardening is not taken into consideration, then the corresponding value is $q^* = \pi c + \gamma h = 3.14 \cdot 0.1 + 0.02 = 0.334$ MPa, which is almost 1.25 times less.

To conclude, it is essential to consider hardening of clayey soil during consolidation for determining the initial critical load. This results in a significantly higher load-bearing capacity and, consequently, greater economising due to reduced dimensions of the foundation.

4. Effect of Rheological Properties of Partially Saturated Soil on Stress Relaxation in Soil Mass

In engineering practice one often comes across cases in which the stress-strain state of a partially saturated mass of clayey soil changes with time due to relaxation processes occurring as a result of the rheological behaviour of soil and the interaction between the soil mass and the structures. For example, when a pile is driven or pushed into a soil medium, the stress-strain state of the surrounding soil mass undergoes a transformation with time, i.e., the stresses in the soil mass experience relaxation until they finally stabilise at a particular value. A similar behaviour is observed during plate load tests and when a probe or measuring device is indented in the soil medium. In all these and other similar cases, it becomes necessary to monitor the quantitative changes that occur in the stress-strain state of the soil mass with time.

Stress relaxation at the base of loaded plate

To solve this problem, it will be assumed that the soil mass experiences hereditary creep. Suppose a force $P(\tau_1)$ has been applied on the surface of a homogeneous, partially saturated soil mass through a rigid plate, resulting in settlement of the plate by $s(\tau_1)$. Subsequently, the plate presses upon a stationary support (Fig. 6.3) through a proving ring of finite stiffness G_l. The time-dependent variation of the force acting on the plate can be determined provided the rheological properties of the soil are known.

Under the action of a time dependent force $P(t)$ settlement of the plate can be determined using similarity of creep kernel according to the following formula:

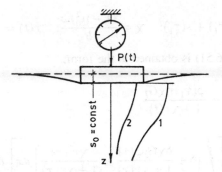

Fig. 6.3. Analytical scheme of stress relaxation in the base of rigid plate.

1—σ_z diagram $(t = 0)$; 2—σ_z diagram $(t = \infty)$.

$$s(t) = \frac{(1 - \mu_0)^2 \omega}{b} \left[\frac{P(t)}{E_m(t)} - \int_{\tau_1}^{t} P(\tau) \frac{\partial}{\partial \tau} \varepsilon(t, \tau) d\tau \right], \qquad \dots (6.47)$$

where b is the width or diameter of the plate and ω is a coefficient that depends on the shape and stiffness of the plate. Further,

$$\left. \begin{aligned} \varepsilon(t, \tau) &= \frac{1}{E_m(t)} + \varphi(\tau) \left[1 - e^{-\delta(t - \tau)} \right]; \\ \varphi(\tau) &= \frac{1}{E_l} + \frac{1}{E_{stab}} \frac{1}{\tau}. \end{aligned} \right\} \qquad \dots (6.48)$$

It is obvious that in the absence of creep and ageing of soil with time, i.e., when $\delta(t, \tau) = 0$ and $E_m(t) = E_0$, the well-known solution to the elastic problem will be obtained from formula (6.47).

The condition describing damping of initial force in the 'dynamometer-plate-soil mass' system can be written as follows:

$$s(\tau_1) + l(\tau_1) = s(t) + l(t) = \text{const.} = \Delta(\tau_1), \qquad \dots (6.49)$$

where

$$l(t) = P(t)/G_l; \quad l(\tau_1) = P(\tau_1)/G_l. \qquad \dots (6.50)$$

Substituting the value of $s(t)$ from formulae (6.47) and $l(t)$ from (6.50) in condition (6.49), the following integral equation in respect of $P(t)$ is obtained:

$$P(t) = \frac{G_l \Delta(\tau_1)}{1 + \chi r(t)} + G_l \chi \int_{\tau_1}^{t} P(\tau) \frac{\partial}{\partial \tau} \frac{\delta(t, \tau)}{1 + \chi r(t)} d\tau, \qquad \dots (6.51)$$

where

$$\Delta(\tau_1) = l(\tau_1) + s(\tau_1); \quad \chi = \frac{(1 - \mu_0^2)\omega}{b}; \quad r(t) = \frac{G_l}{E_m(t)}.$$

The solution of (6.51) is obtained in the form:

$$P(t) = P(\tau_1) \left\{ 1 - \frac{\varphi(\tau_1)\eta\chi G_l}{1 + \chi r(t)} \right.$$

$$\left. \times \int_{\tau_1}^t \exp\left[\int_{\tau_1}^\tau \left[\delta + \frac{\delta\chi G_l\varphi(x)}{1 + \chi r(x)} + \frac{\chi r'(x)}{1 + \chi r(x)} \right] dx \right] d\tau \right\}. \quad \dots \quad (6.52)$$

The above equation determined how the stresses will damp with time in the system 'dynamometer-plate-soil mass', taking into account the finite stiffness of the dynamometer and creep and ageing of the soil skeleton.

As a particular case, when the moduli of the instantaneous strain of the soil are constant and ageing is absent $[E_m(t) = E_m = \text{const.}, E_{\text{stab}} = \infty]$, then $r(t) = r = G_l/E_m$ and $r_1 = G_l/E_l$. Under these conditions, by solving eqn. (6.52), it is found that:

$$P(t) = P(\tau_1) \left[1 - \frac{\chi r_1}{1 + \chi(r + r_1)} \left\{ 1 - \exp\left[-\delta \times \frac{1 + \chi(r + r_1)}{1 + \chi r}(t - \tau_1) \right] \right\} \right]$$
$$\dots \quad (6.53)$$

If the stiffness of the dynamometer is large compared to that of the soil base, then $r = r_1 = \infty$ and, therefore,

$$P(t) = P(t_1) \left[1 - \frac{E_m}{E_m + E_l} \left\{ 1 - \exp\left[-\delta\frac{E_m + E_l}{E_m} \times (t - \tau_1) \right] \right\} \right].$$
$$\dots \quad (6.54)$$

Now, based on the results of relaxation tests conducted in field conditions, the parameters that define the creep of a partially saturated soil base can be easily determined, provided $P(\tau_1)$, $s(\tau_1)$, $P(\infty)$ and $P(t)$ are known, i.e., we have,

$$\left. \begin{aligned} E_m &= \frac{(1 - \mu_0^2)\omega P(\tau_1)}{bs(\tau_1)}; \quad E_l = E_m\frac{P(\infty)}{P(\tau_1) - P(\infty)}; \\ \delta &= -\frac{1}{t}\ln\frac{E_l}{E_m + E_l}\left[1 - \frac{P(t) - P(\infty)}{P(\tau_1)}\frac{E_m + E_l}{E_m}\right]. \end{aligned} \right\} \quad \dots \quad (6.55)$$

Here $1/E_0 = 1/E_m + 1/E_l$. For example, if $E_m = 100$ MPa and $E_l = 10$ MPa, then the initial force will experience relaxation and will reduce 11-fold with the passage of time. Such a wide range of variation of forces in the 'dynamometer-plate-soil mass' system makes it possible to conduct relaxation tests for determining the rheological parameters of the foundations of structures. This considerably reduces the test duration and the time required for conducting field experiments.

A similar approach can be adopted for solving the problem of relaxation in the system 'dynamometer-spherical die-soil mass'. In the case of an absolutely rigid dynamometer

$$P(t) = P(\tau_1) \left[1 - \frac{E_m}{E_m + E_l} \left\{ 1 + \exp \left[-\delta \frac{E_m + E_l}{E_m} \times (t - \tau_1) \right] \right\} \right],$$

$$\dots (6.56)$$

where r_m is radius of the indent produced by the spherical die

$$P(\tau_1) = E_m \frac{4r_m}{3(1 - \mu_0^2)} s(\tau_1).$$

The modulus of instantaneous strain can be determined from the known values of initial force $P(\tau_1)$ and settlement $s(\tau_1)$:

$$E_m = \frac{3(1 - \mu_0^2)}{4r_m} \frac{P(\tau_1)}{s(\tau_1)}. \qquad \dots (6.57)$$

The modulus of long-term deformation can be determined from known values of $P(\tau_1)$ and $P(\infty)$ by the following formula:

$$E_l = E_m \frac{P(\infty)}{P(\tau_1) - P(\infty)}. \qquad \dots (6.58)$$

Formulae (6.55) can be applied to determine the parameters of soil creep.

It may be mentioned that the above solution to the spatial problem of stress relaxation under a spherical die is valid, subject to the condition $s(\tau_1) \leq 0.005 d_{st}$, because in this case the plastic deformation under the die can be neglected [22].

The rheological parameters of partially saturated soil can be determined on the basis of the results of field tests by measuring the contact stresses between the soil and the dynamometer plate indenting in the soil medium.

In the above-mentioned field test, a thin sharp-edged dynamometric plate of rectangular section and length greater than the cross-sectional dimensions is indented in the soil mass at a certain depth, through the shaft wall or the borehole face. The normal contact pressure thus measured represents the sum of stresses, consisting of natural (initial) stresses at the given point of the soil mass and the excess stress that is produced due to indentation of the dynamometric plate, i.e., $\sigma_{dim} = \sigma_n + \sigma_s$.

It is evident that in order to distinguish between the natural and residual stresses, it is necessary to solve the problem of interaction between the dynamometric plate and surrounding soil mass, taking the rheological properties into account. On the one hand, this helps in determining the stress-strain state at the given point of the soil mass and, on the other, the rheological parameters of soil at the given depth. It is especially important to separate the residual component of the excess stress.

In general, the relaxation of excess stress at the 'plate-soil' interface can be defined by the following relation:

$$\left.\begin{array}{l} \sigma_s(t) = \sigma_s(\tau_1)f(t); \\ \sigma_s(\infty) = \sigma_s(\tau_1)f(\infty), \end{array}\right\} \qquad \ldots (6.59)$$

where $\sigma_s(\tau_1)$ is initial residual stress and $f(t)$ is the rate of damping of the residual stresses.

The value of the damping function at the time of stabilisation of the stresses is known as the coefficient of relaxation of the soil. According to the theory of hereditary creep, this coefficient can be determined in the simplest manner by the following relation:

$$K_{rel} = E_0/E_m, \qquad \ldots (6.60)$$

where,

$$\frac{1}{E_0} = \frac{1}{E_m} + \frac{1}{E_l}.$$

It may be mentioned that the coefficient of relaxation is a property of the medium for given physical properties of soil, viz., density and moisture content.

The results obtained by these authors jointly with N.Kh. Kyatov from the solution of the problem of indentation of a dynamometric plate in partially saturated soil having hereditary creep under various indentation conditions (speeds) are described below. From the solution of the problem of expansion of a plane dimensionless crack of width $2b$, it follows that if a uniform pressure is applied on the contour of the crack, the latter is transformed into an elliptical cavity, described by the following equation:

$$y^2/b^2 + u^2/h^2 = 1, \qquad \ldots (6.61)$$

wnere b and h are the semi-axes of the ellipse and u is deflection at a distance $|y| \le b$.

The intensity of pressure is related with parameters of soil deformability as follows:

$$\sigma_s(\tau_1) = E_0 h/[2(1 - \mu_0^2)b]. \qquad \ldots (6.62)$$

If the cross-section of the plate is taken as an elongated ellipse, then the solution to (6.62) can be employed in determining contact pressure between the dynamometric plate and soil, assuming $2b$ to be the plate width and $2h$ its thickness. However, the front portion of the plate in the longitudinal direction, i.e., in the direction of indentation, has the shape of a flat wedge with an elliptical cross-section. It is indented in the soil mass at a particular speed. In this case, the speed at which the soil spreads around is determined by the wedge angle of the front portion of the plate and is described by the relation $\dot{u}(t) = v\,\mathrm{tg}\,\alpha$ (where v is the speed of indentation). The displacement in time t will be $u(t) = vt\cdot\mathrm{tg}\,\alpha$. The time during which the maximum displacement is attained depends on the length of the tapered portion of the plate, i.e., $t_0 = L/v$. Now, on the basis of eqn. (6.62), we have:

$$\sigma(t_0) = \frac{E_0}{2(1-\mu_0^2)}\frac{vt\cdot\mathrm{tg}\,\alpha}{b}. \qquad \ldots (6.63)$$

If the soil displays hereditary creep behaviour described by eqn. (3.28), then the contact stresses will experience relaxation during and after indentation. Therefore, different contact stresses $\sigma(t_0)$ will be set up, depending on the indentation speed. Correspondingly, the nature of damping of the stresses will also differ.

If indentation of the plate is carried out rapidly, the coefficient of stress relaxation will be described by the following expression:

$$\frac{\sigma(t)}{\sigma_m} = \frac{E_0}{E_m} + \left(1 - \frac{E_0}{E_m}\right)\exp\left(-\delta\frac{E_m}{E_0}t\right), \qquad \ldots (6.64)$$

where σ_m is the stipulated instantaneous contact stress.

If indentation is carried out at a slow speed v, then at $t > t_0$ the following expression is obtained:

$$\sigma(t) = \frac{E_0 h v}{2(1-\mu_0^2)bL}\left\{t - \frac{1}{2}\left(1 - \frac{E_0}{E_m}\right) \times \left[\exp\left(-\delta\frac{E_m}{E_0}t\right) - 1\right]\right\}. \qquad \ldots (6.65)$$

As the dynamometric plate measures the residual and natural stresses simultaneously, it is desirable to distinguish between the two. For this purpose, the dynamometric plate is indented up to different depths H_1 and H_2 from the surface of the soil mass both horizontally as well as vertically. The residual pressure will not depend solely on the rheological properties of soil:

$$\left.\begin{array}{ll}\sigma_{\mathrm{dim}}^h(H_1) = \xi_0\gamma H_1 + \sigma_s; & \sigma_{\mathrm{dim}}^h(H_2) = \xi_0\gamma H_2 + \sigma_s; \\ \sigma_{\mathrm{dim}}^v(H_1) = \gamma H_1 + \sigma_s; & \sigma_{\mathrm{dim}}^h(H_2) = \gamma H_2 + \sigma_s,\end{array}\right\} \qquad \ldots (6.66)$$

A comparison of the measurements yields the following possible equations for determining the coefficient of lateral pressure at rest:

$$\left.\begin{array}{l}\xi_0 = [\sigma_{\mathrm{dim}}^h(H_2) - \sigma_{\mathrm{dim}}(H_1)]/[\gamma(H_2 - H_1)]; \\ \xi_0 = 1 - (\sigma_{\mathrm{dim}}^v - \sigma_s)/(\gamma H).\end{array}\right\} \qquad \ldots (6.67)$$

It follows from the above that the excess stresses do not affect the method of determination of the coefficient of lateral pressure. Therefore, the excess pressures may be determined on the basis of eqns. (6.67):

$$\left.\begin{array}{l} \sigma_s = \sigma_{\text{dim}}^h - \xi_0 \gamma H; \\ \sigma_s = \sigma_{\text{dim}}^v - \gamma H. \end{array}\right\} \qquad \dots (6.68)$$

On the basis of the solutions described above and the results of measurements of contact stresses under rapid indentation of the dynamometric plate, the moduli of total and instantaneous deformation of soil can be determined by the following relations:

$$\left.\begin{array}{l} E_0 = 2\sigma_s(\infty) b(1 - \mu_0^2)/h; \\ E_m = 2\sigma_s(0) b(1 - \mu_0^2)/h, \end{array}\right\} \qquad \dots (6.69)$$

The parameter of soil creep can be determined by the expression:

$$\delta = -\frac{1}{t} \frac{\sigma_s(\infty)}{\sigma_s(0)} \ln \frac{\sigma_s(t) - \sigma_s(\infty)}{\sigma_s(0) - \sigma_s(\infty)}, \qquad \dots (6.70)$$

where $\sigma_s(\infty)$ and $\sigma_s(0)$ are the final and initial excess stresses respectively.

The formulae for determining soil parameters under other conditions of indentation of dynamometric plate can be obtained in a similar manner.

Stress relaxation around pile

It is a well-known experimentally established fact that after a pile is driven in soil, the surrounding soil mass experiences excess stresses over and above the natural (initial) stresses. These stresses undergo relaxation with time, which is accompanied by a gradual reduction of the load-bearing capacity. This phenomenon, known as pseudo resistance, is observed in partially saturated clayey, sandy and frozen soils.

Suppose that an elastic cylindrical body (pile) of length l initial radius $r_1 \ll l$ is indented in a partially saturated soil medium at time τ_1. It is required to determine the stress-strain state of the soil mass in space and time, keeping in mind that the excess stresses experience relaxation after the pile is driven in.

The initial contact stress on the pile surface is determined from the condition of plastic flow of soil:

$$\sigma_1(\tau_1) = \left[\frac{n}{m-1} + \frac{2m\sigma_2\tau_2^2 - n(r_2^2 - \rho^2)}{r_2^2 - \rho^2 + m(r_2^2 - \rho^2)} \right] \times \left(\frac{r_1}{\rho} \right)^{\frac{1-m}{n}} - \frac{n}{m-1},$$

$$\dots (6.71)$$

where ρ is the radius of the plastic zone, σ_2 is the stress on the external surface of the soil cylinder, $m = \text{tg}(\pi/4 + \varphi/2) \, \text{ctg}(\pi/4 - \varphi/2)c$ and $n = 2c \cdot \text{ctg}(\pi/4 + \varphi/2)$.

If $r_2 \to \infty$, i.e., if the zone influenced by jacking of the pile is sufficiently large, then

$$\sigma_2(\tau_1) = \left(\frac{n}{m-1} + \frac{rm\sigma_2 + n}{1+m} \right) \left(\frac{r_1}{\rho} \right)^{\frac{1-m}{n}} - \frac{n}{m-1},$$

If, in addition the indenter geometry is such that $\rho \approx r_1 = r_p$, then

$$\sigma_1(\tau_1) = \frac{2m\sigma_2 + n}{1+m},$$

at $\sigma_2 = 0$ and $\varphi = 0$, $\sigma_1(\tau_1) = c$, i.e., the contact stress is equal to the soil cohesion.

The solution of (6.71) describes the relation between contact stresses $\sigma_1(\tau_1)$, pile radius r_1, radius of the zone of influence r_2, initial stress in soil σ_2 and the strength properties of the soil.

The time-dependent contact stress is determined from the following integral equation:

$$\sigma_1(t) = \sigma_1(\tau_1) \left[1 + \frac{\dot{\sigma}_1(\tau_1)}{\sigma_1(\tau_1)} \int\limits_{\tau_1}^{t} \exp\left\{ -\int\limits_{\tau_1}^{\tau} \delta \left[1 + \frac{E_c \varphi(x)}{\beta(A-1)} \right] dx \right\} d\tau \right],$$

$$\ldots (6.72)$$

wherein for plane stress condition:

$$A = \frac{\beta[r_2(1-\mu_0) - r_1(1+\mu_0)]}{(r_1 + r_2)(1 - \mu_p)},$$

while for plane strain condition,

$$A = \frac{\beta[r_2(1+\mu_0) - r_1(1 - 2\mu_0^2 - \mu_0)]}{(r_1 + r_2)(1 - 2\mu_p^2 - \mu_p)};$$

$$\beta = E_p / E_m. \qquad \ldots (6.73)$$

If $\varphi(\tau) = 1/E_l = \text{const.}$ and $\alpha = E_p / E_l$, then:

$$\sigma_1(t) = \sigma_1(\tau_1) \left[1 + \frac{\dot{\sigma}_1(\tau_1)\beta(A-1)}{\sigma_1(\tau_1)\delta[\beta(A-1) + \alpha A]} \right. $$
$$\left. \times \left\{ 1 - \exp\left[-\delta \frac{\beta(A-1) - A}{\rho(A-1)} \right] \right\} \right].$$

From the foregoing, the residual stresses at $t \to \infty$ can be easily determined as follows:

$$\sigma_1(\infty) = \sigma_1(\tau_1) \left\{ 1 + \frac{\dot{\sigma}(\tau_1)}{\sigma_1(\tau_1)} \frac{\beta(A-1)}{\delta[\beta(A-1) + \alpha A]} \right\}. \qquad \ldots (6.74)$$

The initial contact stress can be determined from formula (6.71), while the rate at which it changes at the initial moment can be determined on the basis of eqn. (6.72), i.e.,

$$\dot{\sigma}_1(\tau_1) = \frac{\sigma_2 B\delta}{A-1} - \sigma_1(\tau_1)\varphi(\tau_1)\delta\frac{E_p}{\beta}\frac{A}{A-1}, \qquad \ldots \ (6.75)$$

where,

$$B = \frac{E_c\varphi(\tau_1)[\tau_1 r_2(1-\mu_0) - r_2^2(1+\mu_0)]}{r_1(r_1+r_2)(1-\mu_p)}.$$

If the rigidity of the pile material is greater than that of soil, i.e., when $\beta \to \infty$

$$\sigma_1(\infty) = \sigma_1(\tau_1)E_l(E_m + E_l).$$

If it is assumed that $E_m = 50$ MPa and $E_l = 5$ MPa, then it is found that the initial stress around the rigid pile reduces to one-tenth.

7

Consolidation and Creep of Saturated Clayey Soils

1. General Principles

The soil mass containing saturated layers and interacting with the foundation of the structure is referred to as a saturated base. In these bases, pore pressure resistance is invariably produced under the action of an external load. This pore pressure dissipates, its rate of dissipation depending on numerous factors, including the rheological properties of soil skeleton, coefficient of permeability and compressibility of pore water, dimensions of the soil mass being consolidated and so forth. In view of these, the temporal behaviour of saturated bases is governed, on the one hand, by consolidation and, on the other, by creep of the soil skeleton. Differentiation of the process of soil consolidation into consolidation due to seepage and secondary consolidation is merely formal. Secondary consolidation is related to strain under compressive consolidation, which has been observed in laboratory experiments to grow proportionately with the logarithm of time after almost total dissipation of pore pressure. However, many researchers seem to forget that the rate of deformation of the skeleton during secondary consolidation is very small; the skeleton encounters no resistance from pore water and its deformation can increase at almost negligible pore pressure gradients. In other words, the mutual penetration of soil skeleton and pore water at negligibly small rates produces no significant pressure gradient.

Secondary consolidation may manifest itself in foundations of structures in various ways depending on the width of the foundation and thickness of the layer of clay being consolidated. If the latter is small compared to the width of the foundation, then consolidation is akin to constrained compression. However, if the thickness of the layer of clay is comparable with the width of the foundation, then, besides the other factors, the increase in settlement with time also becomes a function of the skeleton creep, which is not affected much by consolidation through drainage.

In all cases, the consolidation process is accompanied by mutual displacement of the soil skeleton and pore water. However, the magnitude of displacement differs during the various stages of consolidation, thereby producing different levels of resistances in pore water.

Quantitative estimation of the geomechanical process in a saturated soil base depends, on the one hand, on the solution to the problem of consolidation to determine the spatial and temporal distribution of stresses between the soil skeleton and pore water and, on the other, on determination of soil deformation, taking into account the skeleton creep under variable load conditions arising due to the consolidation process.

Depending on the selected rheological model of soil skeleton and compressibility of pore water, one can arrive at different laws of distribution of stresses between the skeleton and pore water and, consequently, different laws for describing the settlement of base with time. This explains why researchers apply different rheological models to describe creep deformation of the soil skeleton, the choice falling on the model that provides the best accord between the particular analytical solution and the experimental results and field observations.

Under one-dimensional consolidation, the solution to the problem of consolidation and creep can be obtained theoretically as well as experimentally without much difficulty. However, the solutions to the two- and three-dimensional problems cannot be obtained with that degree of accuracy and reliability required for practical purposes. This is due to the fact that in two- and three-dimensional problems, the soil skeleton experiences both consolidation strain and shearing strain. The former is directly related to the consolidation process but the latter develops independently. In view of the foregoing, it becomes necessary to redefine the concept of degree of consolidation of foundations of structures for two- and three-dimensional problems.

The concept of degree of consolidation was introduced for determining the settlement of soil layer with time under one-dimensional consolidation, based on the theory of consolidation by drainage. In this concept, the degree of consolidation was related to the dissipation of pore pressure in the layer of saturated soil whose skeleton is incapable of creep. If the soil skeleton displays a distinct creep behaviour, then the concept of degree of consolidation of such soil cannot be related solely to dissipation of pore pressure, because the time of completion of dissipation does not coincide with the time of completion (stabilisation) of the consolidation process.

The estimation of settlement of the base with time as a function of the degree of consolidation is further complicated in the case of two- and three-dimensional problems because the settlement of the base occurs due to consolidation strain as well as shearing of the soil:

$$s(t) = s_v(t) + s_\gamma(t), \qquad \qquad \dots (7.1)$$

where $s_v(t)$ is settlement of the base due to volumetric strain or consolidation process and $s_\gamma(t)$ is settlement of the base due to shearing creep of the soil.

In view of the above, it is necessary to introduce two time-dependent functions in order to evaluate the degree of completion of the total settlement (if this is possible in the rheological sense):

$$s(t) = s_v(\infty)U_v(t) + s_\gamma(\infty)U_\gamma(t), \qquad \dots \ (7.2)$$

where $s_v(\infty)$ and $s_\gamma(\infty)$ are stipulated stabilised settlement due to consolidation strain and shearing respectively; $U_v(t)$ and $U_\gamma(t)$ are time-dependent functions that determine the degree of completion of the settlement process due to consolidation and shearing creep respectively.

These functions acquire a particular meaning because the commonly used form of representation of degree of consolidation does not represent the development of settlement with time, i.e., $U(t) \neq U_v(t) + U_\gamma(t)$. It is clear from the above discussion that there is a need to improve the existing methods of estimation of the time-dependent settlement of base according to the theory of consolidation by drainage using the concept of degree of consolidation. Firstly, the whole stabilised settlement should not be multiplied by the degree of consolidation, i.e., $s(t) \neq s(\infty)U(t)$. Secondly, the settlement of base with time should be determined by taking into account the volumetric and shearing strain of the skeleton, i.e., with the help of functions that govern the time-dependent process:

$$s(t) = s_v(t) + s_\gamma(t) = s_v(\infty)U_v(t) + s_\gamma(\infty)U_\gamma(t), \qquad \dots \ (7.3)$$

where s_γ and s_v represent stabilised settlement due to shearing and volumetric strains, which are determined by the theory of linear elastic medium using formulae (6.13) and (6.14) respectively.

Hence, the unconventional methods of the theory of consolidation by drainage should be applied for analysis and prediction of the settlement of bases in two- and three-dimensional problems of consolidation and creep of soil skeleton.

2. Basic Equations of Consolidation and Creep of Saturated Soils

Estimation of the settlement of foundations of structures with time constitutes one of the fundamental problems of applied geomechanics in civil engineering. At present, this is done by applying the theory of consolidation and creep of multiphase soil consisting of a viscoelastic skeleton and pore water with dissolved gases. In such soils the formation of stress and strain fields occurs in a peculiar manner that significantly differs from similar phenomena in single-phase soil. The main distinguishing feature is that in saturated soil the development and transformation of the stress-strain state is spatial by nature and depends on the rheological properties of the soil skeleton, deformation and seepage characteristics of pore water, dimensions of the soil mass being consolidated etc. The duration of the process of deformation of the soil may vary from a few days to a few decades.

The changes that occur in the stress-strain state of saturated bases can be divided into three stages: initial, intermediate and final.

During the initial stage of the stress-strain state of saturated soils there is no change in the relative proportions of the phases in a unit volume of soil, but there is intense interaction between the phases, resulting in redistribution of the total stresses between the skeleton and pore water. The duration of this stage depends mainly on the rheological properties of the soil skeleton under volumetric deformation. The stress-strain state of soil mass in the initial stage can be described by the equations of the rheological state of the skeleton and pore water. During this stage drainage can be ignored. However, one cannot rule out the possibility of redistribution of stresses in pore water and the skeleton due to flow of pore water from a high-pressure region to a low-pressure region (without taking into consideration the effect of the boundary conditions of free drainage). The initial stage ends with total redistribution of the external load between the skeleton and pore water and extends into the intermediate stage.

On the one hand, the intermediate stress-strain stage of a saturated base is distinguished by further redistribution of stresses between the skeleton and pore water and intensive change in their proportion in a unit volume, resulting in development of consolidation strain. On the other hand, the intermediate stage is accompanied by development of shearing creep under the effect of shearing stresses and total effective stresses that vary with time.

The duration of the intermediate stage depends on numerous factors, of which the leading ones are width of foundation that determines the depth of the active zone of consolidation, rheological properties of the soil skeleton and the coefficient of permeability and compressibility of pore water. The intermediate stage of the stress-strain state can be described by the equations of the theory of consolidation of multiphase soil with consideration of volumetric creep of the skeleton and the equations of shearing creep with appropriate boundary and initial conditions.

The extent to which the creep behaviour of the soil skeleton affects the duration of the intermediate stage can be evaluated by means of the following non-dimensional parameter:

$$\mu_c = \delta_v (H_a^2 / c_v), \qquad \qquad \dots (7.4)$$

where δ_v is creep of soil skeleton under volumetric strain, H_a is the depth of active zone of consolidation and C_v is the coefficient of consolidation.

For soils with $k \geq 10^{-5}$ cm/s, at $\mu_c \leq 0.001$, the duration of the intermediate stage is determined by the skeleton creep and, therefore, there is no need to examine the consolidation problem. For soils with $k \leq 10^{-8}$ cm/s, at $\mu_c \geq 10$, the duration of the intermediate stage is determined exclusively by the seepage behaviour and, therefore, it is not essential to take into account the creep phenomenon.

Hence, it is first necessary to determine μ_c in order to decide whether the effect of a particular factor on the duration of the intermediate stage needs to be considered in the given case.

The intermediate stage ends with the dissipation of excess pore pressure and total transfer of the external load to the soil skeleton. However, this does not mean that the process of growth of skeleton strain is also over. Because of the rheological behaviour of the skeleton, the deformation may continue for a considerable period after the stresses have stabilised. The growth of deformation with time proceeds at a continuously decreasing or steady rate, depending on the shearing and volumetric viscosity of the soil skeleton. In view of the foregoing, there is a possibility of continuation of the consolidation process and its extension into the final stage.

The final stage of the stress-strain state of saturated base is distinguished by an absence of pore pressure and further increase in skeleton creep. The duration of this stage is determined exclusively by the rheological properties of the skeleton and it may either decrease with time or remain steady. The stress-strain state in the final stage can be described by the rheological equations of state of the soil skeleton.

For the sake of clarity of the discussion to follow, the initial equations required for describing the stress-strain state of saturated soil in space and time are given below.

Equilibrium equations:

$$\left.\begin{array}{l} \dfrac{\partial \sigma_x}{\partial x} + \dfrac{\partial \tau_{xy}}{\partial y} + \dfrac{\partial \tau_{xz}}{\partial z} = X - \dfrac{\partial u_w}{\partial x}; \\ \dotfill \end{array}\right\} \qquad \dots (7.5)[1]$$

where u_w is pore pressure.

Geometrical equations:

$$\left.\begin{array}{l} \varepsilon_x = \dfrac{\partial u}{\partial x}; \quad \varepsilon_{xy} = \dfrac{\partial u}{\partial y} + \dfrac{\partial v}{\partial x}; \\ \dotfill \end{array}\right\} \qquad \dots (7.6)[2]$$

Rheological equations for soil skeleton:

$$\left.\begin{array}{l} \varepsilon_i(t) = 2\psi_i^{(0)} \sigma_i(t) + 2\Phi_i[\psi_i \sigma_i(t)]; \\ \varepsilon_v(t) = \psi_v^{(0)} \sigma_v(t) + \Phi_v[\psi_v \sigma_v(t)], \end{array}\right\} \qquad \dots (7.7)$$

---

[1]The rest of the equations may be obtained by substitution of arguments x, y, z in a cyclic order.

[2]*Ibid.*

126

where $\Phi_i[y(t)]$ and $\Phi_v[y(t)]$ are Volter integral operators with kernels $K_i(t, \tau)$ and $K_v(t, \tau)$ that describe the creep of soil skeleton under shearing and all-round compression respectively; $\psi_i^{(0)}$, $\psi_v^{(0)}$ and ψ_i, ψ_v are functions that describe the mechanical properties of the soil skeleton at the commencement and end of the time duration under consideration:

$$\left.\begin{aligned}
\psi_i^{(0)} &= \varepsilon_i^{(0)}(\sigma, \sigma_i, \lambda_\sigma, K_\sigma)/(2\sigma_i); \\
\psi_v^{(0)} &= \varepsilon_v^{(0)}(\sigma, \sigma_i, \lambda_\sigma, K_\sigma)/\sigma_v; \\
\psi_i &= \varepsilon_i(\sigma, \sigma_i, \lambda_\sigma, K_\sigma)/(2\sigma_i); \\
\psi_v &= \varepsilon_v(\sigma, \sigma_i, \lambda_\sigma, K_\sigma)/\sigma_v.
\end{aligned}\right\} \qquad \dots (7.8)$$

If the type of stressed state λ_σ and the loading trajectory parameter K_σ do not effect the form of relations (7.8), then the latter can be simplified and written in the well-known form adopted in the theory of small elastoplastic deformations [1, 17].

$$\psi_i = \varepsilon_i(\sigma, \sigma_i)/(2\sigma_i); \quad \psi_v = \varepsilon_v(\sigma)/\sigma_v.$$

In this case, the relation between the stress and strain components can be written as follows:

$$\left.\begin{aligned}
\varepsilon_x &= \psi_i(\sigma_x - \sigma) + \psi_v\sigma; \\
\varepsilon_{xy} &= \psi_i\tau_{xy};
\end{aligned}\right\} \qquad \dots (7.9)^3$$

The physical equations for pore water with dissolved gases are obtained by examining two states of pore water and the condition of constancy of the mass of water, bubbles and dissolved gas [17]:

$$\alpha u'_g n'_{w,g} + \rho'_g n'g = \alpha u''_g n''_{w,g} + \rho''_g n''g, \qquad \dots (7.10)$$

where α is Henry's solubility coefficient for gas mixture, $n_{w,g}$ is the volume of pore water together with dissolved gas, n_g is the volume of gas bubbles, u_g is the pressure inside gas bubbles and ρ_g is the density of gas mixture (air).

If it is assumed that the volume of pore water in which gas is present in the form of a solution does not vary significantly (i.e., $n'_{w,g} = n''_{w,g} = n_{w,g}$) and that $n_g = n - n_{w,g}$, $\rho_g = xu_g$, then it is found that:

$$\alpha u'_g + u'_g(1 - S'_r/S_r) = \alpha u''_g + u''_g(1 - S''_r)/S''_r, \qquad \dots (7.11)$$

where $S'_r = n_{w,g}/n$.

By assuming $S''_r = 1$, the pressure u_g^* in the gas bubbles at the time of dissolution can be readily determined from the above equation as:

$$u_g^* = u'_g + u'_g \frac{1 - S'_r}{\alpha S'_r}.$$

---

[3] See footnote to eqns.(7.5) and (7.6)

For example, if $S'_r = 0.9$, $u'_g = 0.1$ MPa and $\alpha = 0.0205$, then it is found that $u^*_g = 0.65$ MPa; if $S'_r = 0.8$, $u^*_g = 1.35$ MPa.

Hence, during consolidation of partially saturated soil, the degree of saturation increases with the pore water pressure and may even attain a value equal to unity. This can be demonstrated by representing eqn. (7.11) in the following form:

$$S''_r = \frac{u''_g/u'_g}{(1-\alpha)(u''_g/u'_g - 1) + 1/S'_r}.$$

For example, if $S'_r = 0.9$, $u'_g = 0.1$ MPa and $u''_g = 0.2$ MPa, $S''_r = 0.957$. As the pressure is increased to $u''_g = 0.4$ MPa, $S''_r = 0.987$ and when $u''_g = 0.6$ MPa, $S''_r = 0.997$.

The coefficient of relative compressibility of pore water can be determined on the whole from the increment in degree of saturation and the corresponding change of pressure in gas bubbles Δu_g. Let us consider the ratio $S'_r/S''_r = n''/n' = (n'' - n')/(n' + 1)$. As $(n'' - n')/n' = m_g - \Delta u_g$, it is found that

$$m_{g,w} = (1 - S'_r/S''_r)\Delta u_g. \qquad \ldots (7.12)$$

The coefficient of volumetric compressibility of water with dissolved gas will be given by:

$$K_{g,w} = 3/m_{g,w} = 3\Delta u_g/(1 - S'_r/S''_r).$$

The pore water pressure is related to the pressure in gas bubbles by the relation:

$$u_g = u_{g,w} + 2q/r, \qquad \ldots (7.13)$$

where q is surface tension, r is bubble radius and $u_{g,w}$ is the pressure in water with dissolved air.

Substituting (7.13) in formula (7.12) and neglecting the change in radius of the bubbles, it is found that:

$$\left. \begin{array}{l} m_{g,w} = (1 - S'_r/S''_r)/\Delta u_{g,w}; \\ K_{g,w} = 3\Delta u_{g,w}/(1 - S'_r/S''_r). \end{array} \right\} \qquad \ldots (7.14)$$

In the discussion below, the subscript g, w in the notations $m_{g,w}$, $K_{g,w}$ and $u_{g,w}$, will be replaced by w, but it will continue to be implied that it pertains to water with dissolved and undissolved air.

For saturated soil as a whole, the rheological equation of state is related to the stress-strain state of soil mass in the initial stage of consolidation. By introducing the concepts of reduced modulus of volumetric compressibility and reduced Poisson's ratio for soil as a whole, the volumetric strains of the skeleton and pore water can be correlated by an expression of the type given below:

$$\varepsilon_{sk} = n\varepsilon_w. \qquad \ldots (7.15)$$

Assuming that $\varepsilon_{\text{red}} = \sigma/K_{\text{red}}$, $\varepsilon_{sk} = \sigma'/K_{sk}$, $\varepsilon_w = u_w/K_w$ and $\sigma = \sigma' + u_w$, it is found that:

$$K_{\text{red}} = K_{sk} + K_w/n; \qquad \dots (7.16)$$

$$\mu_{\text{red}} = \frac{K_{\text{red}} - 2G}{2(K_{\text{red}} + G)}. \qquad \dots (7.17)$$

In the above equations $G = G_{\text{red}}$ because pore water does not offer resistance to shearing deformation.

The expressions for K_{red} and μ_{red} are not so simple when the deformability of the skeleton and pore water are non-linear because, in this case, it becomes necessary to solve transcendental equations. However, if the stresses do not vary in a wide range and tangent moduli are used in the above expressions, then eqns. (7.16) and (7.17) can be employed by inducting correction factors for K_{sk} and K_w as σ and u_w are increased (Fig. 7.1).

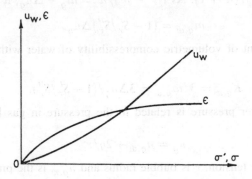

Fig. 7.1. Characteristic curve depicting variation of the pore pressure u_w and volumetric strain ε as functions of effective stress, σ, and total stress $\sigma = (\sigma_1 + \sigma_2 + \sigma_3)/3$ for partially saturated clay in the absence of drainage.

From the solution of simultaneous equations (7.15) and (7.16) it follows that:

$$u_w = \sigma K_w/(nK_{sk} + K_w). \qquad \dots (7.18)$$

Upon changing over to tangent moduli, it is found that:

$$\Delta u_w = \Delta\sigma K_w/(nK_{sk} + K_w); \qquad \dots (7.19)$$

where $1/K_{sk} = 1/K_{sk}^e + 1/K_{sk}^p$.
When the load on the soil is removed $1/K_{sk} = 1/K_{sk}^e$ and therefore,

$$\Delta u_w' = \Delta\sigma K_w/(nK_{sk}^e + K_w).$$

Hence, when a load is first applied on clayey soil and then reduced by $\Delta\sigma$, the excess pore pressure given below is developed in the closed system:

$$\Delta u_{w,\text{res}} = \Delta\sigma K_w \left(\frac{1}{nK_{sk} + K_w} - \frac{1}{nK_{sk}^e + K_w} \right).$$

Upon load removal, the reduced modulus and Poisson's ratio will change, on the whole, in corresponding fashion:

$$\left. \begin{array}{l} K'_{\text{red}} = K_{sk}^e + K_w/n; \\[2mm] \mu_{\text{red}} = \dfrac{K'_{\text{red}} - 2G}{2(K'_{\text{red}} + G)}. \end{array} \right\} \qquad \ldots (7.20)$$

The expression for reduced coefficient of relative compressibility of saturated soil can be written as:

$$m_{v,\text{red}} = n\frac{m_v m_w}{m_v + nm_w}.$$

Obviously, $m_{v,\text{red}} < m_v$.

In the case of load removal, the corresponding expression is obtained as:

$$m'_{v,\text{red}} = n\frac{m_v^e m_w}{m_v^e + nm_w}.$$

As $m_v^e < m_v = m_v^e + m_v^p$, it follows that $m'_{v,\text{red}} < m_{v,\text{red}}$.

Knowing the reduced modulus, the deformation and settlement of a saturated layer under one-dimensional consolidation can be determined from the following relation:

$$s = h\varepsilon_1 = \Delta\sigma m_{v,\text{red}} h.$$

It is possible to examine in a similar manner the problem of distribution of total stresses between the soil skeleton and pore water containing dissolved and trapped gases with consideration of creep, non-linear deformability, dilatation behaviour etc. In all these cases one has to deal with transcendental equations which may be solved with the help of computers.

For analysis of the stress-strain state of soil, it is convenient to employ graphic representation of the $u_w = f(\sigma)$ and $\varepsilon = f(\sigma)$ relations (Fig.7.1).

The equation of consolidation of saturated clayey soil may be obtained from the condition of constancy of the mass of pore water and mineral particles in an elementary volume during consolidation [17, 20]:

$$\frac{\partial\varepsilon}{\partial t} - n\frac{\partial\varepsilon_w}{\partial t} = \frac{K_f}{\gamma_w}\nabla^2 u_w, \qquad \ldots (7.21)$$

where ε and ε_w represent volumetric strain of soil and pore water respectively, n is porosity, k is coefficient of permeability, γ_w is unit weight of pore water and ∇ is the Laplace operator.

The above equation is valid irrespective of the law by which the deformation of the soil skeleton and pore water is described. In the absence of seepage, i.e., at $k = 0$, the condition of compatibility of volumetric strain of the skeleton and pore water ($\varepsilon = n\varepsilon_{sk}$) can be obtained from eqn. (7.21); if the skeleton is incompressible ($\varepsilon = 0$), eqn. (7.21) yields the equation of seepage of compressible liquid into a rigid porous medium; if pore water is assumed to be incompressible ($\varepsilon_w = 0$), the Terzaghi equation of consolidation by drainage is obtained from equation (7.21); finally, when $\varepsilon = \varepsilon_w = 0$, eqn. (7.21) yields the equation of seepage of an incompressible liquid into an incompressible porous medium. Thus, we now have all the equations required for solving the problem of creep and consolidation of saturated soil. It may be added that the Terzaghi condition $\sigma = \sigma' + u_w$ and the initial and boundary conditions of the given problem must be satisfied during the above consolidation process.

In the simplest case, when the soil skeleton is elastic and pore water is linearly compressible, the system of equations for solving the two-dimensional problem of consolidation consists of the equilibrium equations in displacements u and v and the equations of continuity of flow of the liquid and solid phases in the course of soil consolidation:

$$\left.\begin{aligned}
G\nabla^2 u + \frac{G+K}{3}\frac{\partial \varepsilon_v}{\partial x} &= \frac{\partial u_w}{\partial x}; \\
G\nabla^2 w + \frac{G+K}{3}\frac{\partial \varepsilon_v}{\partial z} &= \frac{\partial u_w}{\partial z}; \\
\frac{\partial \varepsilon_v}{\partial t} + \frac{n}{K_w}\frac{\partial u_w}{\partial t} &= \frac{k}{\gamma_w}\nabla^2 u_w.
\end{aligned}\right\} \qquad \dots (7.22)$$

3. Consolidation and Creep of Saturated Clays under One-dimensional Compression

In engineering practice, one-dimensional compression of soil in foundations of structures is quite common. Such conditions are obtained during consolidation of bases of foundations of structures with large dimensions in plan and also during consolidation of a weak layer that constitutes part of a multilayer base.

Let us try to obtain the solution of the one-dimensional problem of consolidation and creep of saturated base, taking into account the various properties of the soil skeleton and pore water. The solution to each problem includes all three stages of the consolidation process.

Consideration of hereditary creep of soil skeleton creep and compressibility of power water

While examining this particular case in Chapter 2, it was pointed out that creep deformation of the soil skeleton due to hardening may be estimated by introducing an additional term in the formula for settlement. This greatly simplifies the solution of the one-dimensional problem of consolidation.

The parameters of the settlement equation may be determined from the results of compression tests conducted in laboratory conditions (see section 4.3).

Consideration of damping skeleton creep and compressibility of pore water

The creep deformation under one-dimensional consolidation can be described by the equation of state from the theory of ageing:

$$\varepsilon(t) = m_{v1}\sigma_1(t) + m_{v3} \int_{t_1}^{t} \frac{\sigma_1(\tau)}{\tau} d\tau, \qquad \dots (7.23)$$

where m_{v1} and m_{v3} are the coefficients of relative elastic and inelastic compressibility respectively.

When the load is constant with time, i.c., $\sigma_1(t) = \sigma_1 = \text{const.}$, the above equation becomes modified to:

$$\varepsilon_1(t) = \sigma_1[m_{v1} + m_{v3}\ln(t/t_1)].$$

Considering eqn. (7.23) simultaneously with the equation of consolidation (7.21) and assuming the pore water to be linearly compressible, it is found that:

$$(1 + A_w)\frac{\partial u_w}{\partial T} - A_h\frac{q - u_w}{T} = \frac{\partial^2 u_w}{\partial \xi^2}, \qquad \dots (7.24)$$

where $A_w = nm_w/m_{v1}$; $A_h = m_{v3}/m_{v1}$; $c_v = k/(\gamma_w/m_{v1})$; $T = c_v t/h^2$; $\xi = z/h$.

The initial condition may be determined from the stipulated instantaneous deformation of the soil skeleton or from the condition of prolonged deformation of the skeleton:
—in the first case

$$u_w(z, \tau_1) = q(z)A_0, \qquad \dots (7.25)$$

where $A_0 = m_{v1}/(m_{v1} + nm_w)$;
—in the second case

$$u_w(z, t) = q(z)[1 - (1 - A_0)(t/\tau_1)^{-\lambda}], \qquad \dots (7.26)$$

where $\lambda = m_{v3}/(m_{v1} + nm_w)$.

It is evident from formula (7.26) that at $t = \tau_1$, the usual initial condition $u_w(\tau_1) = qA_0$ is obtained, whereas at $t \to \infty u_w \to q$.

Let us examine the solution (7.24) for the initial condition (7.25) and boundary conditions $u_w(0, t) = 0$ and $\partial u_w(n, t)/\partial z = 0$. At $q(z) = q = \text{const.}$, it is found [17] that:

$$u_w(\xi, T) = \frac{4q}{\pi} \sum_{n=1,2,\ldots}^{\infty} \frac{1}{n} \left(\sin \frac{\pi n}{2} \xi \right) F_n(T), \qquad \ldots \ (7.27)$$

where

$$F_n(T) = (A_0 - 1) \left(\frac{T_1}{T} \right) \exp[-B_n(T - T_1)] + 1 - \frac{B_n T}{c+1} + \frac{B_n^2 T^2}{(c+1)(c+2)}$$

$$- \frac{B_n^3 T^3}{(c+1)(c+2)(c+3)} + \cdots;$$

$$B_n = \frac{(\pi n)^2}{h^2(1 + A_w)}; \quad c = \frac{A_h}{1 + A_w}.$$

The degree of consolidation can be evaluated from the following expression:

$$U_0(T) = 1 - \frac{8}{\pi^2} \sum_{n=1,3,\ldots}^{\infty} \frac{1}{n^2} [F_n(T) + G_n(T)], \qquad \ldots \ (7.28)$$

where

$$G_n(T) = \frac{A_0 - 1}{A_0} \left\{ 1 - \left(\frac{T_1}{T} \right)^c \exp[-B_n(T - T_1)] - B_n T_1 \right.$$

$$\times \exp(B_n T_1)\phi(B_n T, c) \Big\} + A_h \left[\ln \frac{T}{T_1} - B_n \frac{T}{c+1} + \frac{B_n T^2}{2(c+1)(c+2)} + \cdots \right];$$

wherein $\phi(B_n T, c)$ is partial gamma function of argument $B_n T$ with parameter c.

Expression (7.28) represents the ratio of time-dependent settlement to the settlement that builds up during consolidation by drainage due to the elastic properties of the soil skeleton, i.e., $U_0(T) = s(t)/s_f$. In view of the above, $U_0(T)$ varies from 0 to 1 to attain a value of unity at $t = t_f$ and continues to increase further as the creep strain increases proportionately with the logarithm of time.

Expressions (7.27) and (7.28) include parameters m_{v1}, m_{v3}, m_w and τ_1 which can be readily determined from the results of compression tests on saturated soil specimens under stepped loading (see section 4.3).

An analysis of eqns. (7.27) and (7.28) on the basis of the solution of numerical examples on the computer [17] revealed that when damping skeleton creep is considered in the form of relation (7.23), it leads to new results that differ qualitatively from those obtained by other theories of consolidation. Firstly, these results confirm the experimentally established fact (Fig. 7.2) that the inflection point on the settlement vs logarithm of time curve represents the instant of time when the pore pressure is completely dissipated. Secondly, according to this theory, settlement continues to increase proportionately with logarithm of time even after complete dissipation of pore pressure. Thirdly, the relative settlement of layer decreases as the thickness of the layer being consolidated increases.

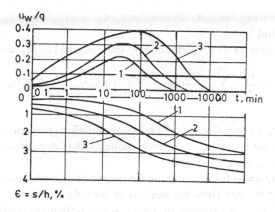

Fig. 7.2. Consolidation and creep curves for identical saturated loam specimens ($l_0 = 0.6$) of different heights tested on compression set up with flexible walls under load $q = 0.2 - 0.4$ MPa.

1—$H = 20$ mm; 2—$H = 74$ mm; 3—$H = 184$ mm.

All the above conclusions have considerable theoretical and practical significance. The last conclusion is especially important as it explains the reasons for reduction of settlement with increase in thickness of the foundation of the structure. This effect can be explained as follows. With the increase in thickness of the consolidated layer, the dissipation of pore pressure through the soil mass slows down, thereby reducing the rate of loading of the soil skeleton. This results in a greater degree of hardening, manifested in reduced level of deformation. The hardening ability of soil is reflected in eqn. (7.23) and may be attributed to restoration and hardening of bonds between particles and their aggregates after application of an external load.

Consideration of hereditary creep and ageing of soil skeleton
An analysis of the solutions presented above revealed that the consolidation process is significantly affected by creep and ageing of the skeleton. By simultaneous consideration of these factors it is possible to describe both the extremal nature of pore pressure variation as well as the logarithmic nature of the growth of settlement with time after complete dissipation of pore pressure.

The solution to the above problem can be obtained by concomitantly examining equation (7.21), representing one-dimensional consolidation, and (7.29), representing hereditary creep of soil skeleton with consideration of ageing. The following differential equation is obtained:

$$(1 + A_w)\frac{\partial^2 u_w}{\partial T^2} + \delta \left[(i - A_w + A_l)\frac{h^2}{c_v} + \frac{A_h}{T} \right] \frac{\partial u_w}{\partial T} = \frac{\partial^3 \dot{v}_w}{\partial \xi^2 \partial T} + \delta \frac{h^2}{c_v} \frac{\partial^2 u_w}{\partial \xi^2},$$

$$\ldots (7.29)$$

where $A_w = nm_w/m_v$, $A_l = m_{v2}/m_{v1}$, $A_h = m_{v3}/m_{v1}$; $c_v = k/(\gamma_w m_{v1})$; $T = c_v t/h^2$ and $\xi = z/h$.

The solution for (7.29) obtained by the author [17, 21] was as follows:

$$u_w(\xi, T) = \frac{4q}{\pi} \sum_{n=1,3,\ldots}^{\infty} \frac{T^2}{n} \sin \frac{\pi n \xi}{2} [C_1 F(\alpha, \gamma, x) + C_2 G(\alpha, \gamma, x)] x e^{\lambda} n^T,$$

$$\ldots (7.30)$$

where $F(\alpha, \gamma, x)$ and $G(\alpha, \gamma, x)$ are degenerating hypergeometric functions of first and second types; α, γ, x are exponents that depend on soil parameters m_{v1}, m_{v2}, m_{v3}, m_w, τ_1, k and c_v.

The expressions for determining the degree of consolidation have been given in [17]. It can be seen from the analysis of formula (7.30) that variation of pore pressure with time is extremal in nature, while settlement continues to increase proportionately with the logarithm of time even after complete dissipation of pore pressure. In addition, the relative settlement vs time curves on a semi-log plot for layers of various thicknesses tend towards different asymptotes, i.e., relative settlement is a variable quantity. For verification of the theoretical conclusions, experiments were conducted on a compression set-up with flexible walls on which identical beam specimens of cross-sectional area 40 cm^2 but different height were tested. It was established by these studies that the maximum pore pressure and the period during which it was attained increased, whereas relative settlement decreased with an increase in height of the specimen.

In the absence of drainage, we have:

$$u_w(t) = u_w(\tau_1) \left\{ 1 - \frac{\delta \gamma(\tau_1)(1 - A_0)}{m_{v1}} \int_{\tau_1}^{t} \exp\left[-A \times (\tau - \tau_1) - B \ln \frac{\tau}{\tau_1} \right] d\tau \right\},$$

$$\ldots (7.31)$$

where

$$u_w(\tau_1) = q\frac{m_v}{m_v + nm_w};$$

$$A = \delta\frac{m_{v1} + m_{v2} + nm_w}{m_{v1} + m_{v2}}; \quad B = \delta\frac{m_{v3}}{m_{v1} + nm_w};$$

$$\varphi(\tau_1) = m_{v2} + m_{v3}/\tau_1.$$

In particular, when $m_{v3} = 0$, it is found that:

$$u_w(t) = u_w(\tau_1) \left[1 - \frac{\delta m_{v2}(1 - A_0)}{A m_{v1}} \{1 - \exp[-A(t - \tau_1)]\} \right].$$

An analysis of the above solutions reveals that in the absence of drainage, pore pressure increases with time up to a certain level, and then stabilises at a value which is always less than the applied load. This conclusion accords well with the experimental results and also the recorded values of pore pressure in cores of earth dams and foundations of structures in the initial stage of consolidation.

Consideration of non-linear deformability and damping creep of soil skeleton
The initial equations can be written in the following form:

$$\delta\varepsilon_1/\delta t = be^{-a\sigma}/t; \quad k = k_0 e^{-a\sigma}; \quad m_w = (1 - S_r)/p_a = \text{const}.$$

For the above initial equation, the solution to the one-dimensional problem is obtained as [21]:

$$\left.\begin{array}{l} u_w(z,t) = \dfrac{1}{a}\ln\left[\dfrac{4 - e^{-aq}}{\pi}\displaystyle\sum_{n-1,3,\ldots}^{\infty}\dfrac{1}{n}\sin\dfrac{\pi nz}{2h} \times F_n(t) + e^{-aq}\right] + q; \\[4mm] s(t) = bhe^{-aq}\left[\dfrac{8}{\pi^2}\displaystyle\sum_{n=1,3,\ldots}^{\infty}\dfrac{1}{n^2}G_n(t) + \ln\dfrac{t}{\tau_1}\right], \end{array}\right\}$$

$$\ldots (7.32)$$

where $F_n(t)$ and $G_n(t)$ are time-dependent functions that also depend on the rheological parameters.

From the structures of the above solution it is evident that pore pressure variation is extremal in nature and that the settlement of a layer increases proportionately with the logarithm of time.

On examining the solutions of the one-dimensional problem of consolidation and creep of saturated clayey soil discussed above, it can be seen that skeleton creep and pore water compressibility significantly affect the variation of pore pressure and settlement with time. The pattern of variation differs significantly from that obtained by other theories of consolidation. Nonetheless, when the solutions given above are applied in practical calculations, it is desirable to make a preliminary assessment of the need to consider skeleton creep in the analysis of consolidation according to criteria (7.4).

Consideration of non-linear deformability of soil skeleton and non-linear permeability of soil
The following discussion concerns estimation of the settlement of foundations of structures that are built on loose, highly compressible, saturated soils such as silts and peats. Let us first examine the analytical solution of the problems, assuming that the non-linear variation of the compressibility of the soil skeleton and pore water as also the permeability of the soil are known, i.e.,

$$e = e_0 - b(1 - e^{-a\sigma}); \quad k = k_0 e^{-na\sigma};$$

$$m_w = (1 - S_r)/(u_w + u_a) \approx (1 - S_r)e^{-a\sigma}.$$

For the relations given above, the one-dimensional consolidation at $n = 1$ is described by the following equation:

$$[ab + (1 - S_r)e_{\text{mid}}]\frac{\partial u_w}{\partial t} = c_v^{(0)}\frac{\partial^2 u_w}{\partial z^2} + ac_v^{(0)}\left(\frac{\partial u_w}{\partial z}\right)^2. \qquad \ldots (7.33)$$

For the boundary conditions $z = 0$, $z = h$, $u_w = 0$, $\varphi = 0$ and the initial conditions $t = 0$, $u_w = \beta_0 q$, $\varphi = \exp[-aq(1-\beta_0)] - \exp(-aq) = \beta$, the solution of the above equation yields the following results:

$$u_w(z,t) = \frac{1}{a}\ln\left[\frac{4\beta}{\pi}\sum_{i=1}^{i=\infty}\frac{1}{i}\exp\left(-\frac{c_v^{(0)}i^2\pi^2 t}{mh^2}\right)\right.$$

$$\left. \times \ \sin\frac{i\pi z}{h} + \exp(-aq)\right] + q; \qquad \ldots (7.34)$$

$$s(t) = \frac{bh}{1 + e_0}[1 - \mu(t) - \exp(-aq)], \qquad \ldots (7.35)$$

where

$$m = ab + (1 - S_r)e_{\text{mid}};$$

$$\mu(t) = \frac{8\beta}{\pi^2}\sum_{i=1}^{i=\infty}\frac{1}{i^2}\exp\left(-\frac{c_v^{(0)}i^2\pi^2 t}{mh^2}\right);$$

$$Q(t) = s_t/s_\infty = 1 - \frac{\mu(t)}{1 - \exp(-aq)}.$$

Let us now consider the solution to the problem of one-dimensional consolidation, taking into account non-linear compressibility and permeability, but without averaging the coefficient of porosity of soil. This solution was derived by the author in collaboration with I.I. Demin and A.A. Rakhmanov. Suppose the non-linear deformability and permeability are defined by the following relations:

$$e = e_0 - b(1 - \psi); \quad k = k_0\psi^n, \qquad \ldots (7.36)$$

where $\psi = \exp[-a(q - u_n)]$; a, b and n are coefficients determined experimentally; e_0 and k_0 are the initial values of the coefficient of porosity and coefficient of permeability respectively and q is the consolidating load.

For the relations of non-linear deformability and permeability adopted above, the equation of one-dimensional consolidation due to drainage may be written as follows:

$$\frac{mB + (a + m_w)\psi}{a\psi(A + \psi)}\frac{\partial\psi}{\partial t} = c_v^{(0)}\frac{\partial}{\partial z}\psi^{n-1}\frac{\partial\psi}{\partial z}, \qquad \dots (7.37)$$

where $c_v^{(0)} = k_0 f_0/(\gamma a)$; $A = (1 + e_0 - b)/b$; $B = (e_0 - b)/b$; f_0 is a coefficient for conversion of units and γ_w is the unit weight of water.

Equation (7.34) will be solved for the following initial and boundary conditions:

$$\left.\begin{array}{l}\psi(z, 0) = \exp[-aq(1 - \beta_0)] = \psi_1; \\ \psi(0, t) = \psi(h, t) = \exp(-aq) = \psi_2,\end{array}\right\} \qquad \dots (7.38)$$

where β_0 is the coefficient of pore pressure equal to u_w/q; it can be determined by means of the following integral substitutions:

$$H(\psi) = \int_{\psi_{min}}^{\psi}\frac{m_w B + (a + m_w)\psi}{c_v^{(0)}a\psi(A + \psi)}d\psi; \quad \phi(\psi) = \int_{\psi_{min}}^{\psi}\psi^{n-1}d\psi, \qquad \dots (7.39)$$

where $\psi_{min} = \psi_2$.

With the help of the above substitution, eqn. (7.37) is converted into the quasi-linear form given below with suitably modified initial and boundary conditions:

$$\frac{\partial H}{\partial t} = \frac{\partial^2\phi}{\partial z^2}, \qquad \dots (7.40)$$

For the above equation the extremal principle has been formulated and with its help all the relations of the finite element method have been obtained. A computer program has been developed which makes it possible to solve a wide range of problems dealing with estimation of the settlement of highly compressible clayey soils.

The curve depicting variation of the degree of consolidation $U(t)$ with time (Fig. 7.3) can be described by the expression:

$$U(t) = U_0(t_0) + [1 - U_0(t_0)][1 - \exp(-\lambda t)], \qquad \dots (7.41)$$

where $U_0(t_0)$ is the initial degree of consolidation; λ is a generalised coefficient that depends on the type of soil, initial height of layer, drainage conditions, deformability parameters, permeability etc. and is reciprocal of time.

By means of an iterative computational scheme it was possible to obtain a solution that takes into account the variation in the height of the consolidating layer during the course of consolidation due to drainage. It was established that consideration of the variation in height of the consolidating layer leads to the following simple relation between the degree of consolidation and time.

$$U(t) = U_0(t_0) + \text{tg}\,\theta t(0 \le t \le t_f), \qquad \dots (7.42)$$

138

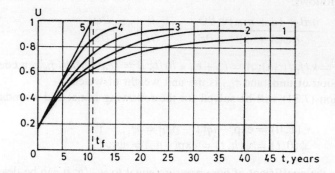

Fig. 7.3. Curves depicting variation of the degree of consolidation due to drainage with time.

1—without taking into account the variation in height of layer; 2 to 5—with consideration of change in the height of the layer after every 10, 5, 2.5 and 0.04 years respectively.

where $\text{tg}\,\theta = [1 - U_0(t_0)]/t_f$; wherein $t_f = 1/\lambda$ is the time of completion of consolidation due to drainage for variable height of the consolidated layer.

For a layer of height 10 m with the parameters $e_0 = 9.19$, $b = 7.37$, $a = 18.22$ MPa^{-1}, $k_0 = 2.91 \times 10^{-5}$ m/hr, $n = 1.95$, $c_v^{(0)} = 1.597 \times 10^{-4}$ m/hr, $A = 0.383$, $B = 0.217$, $m_w = 10^{-2}$ MPa^{-1}, $q = 0.05$ MPa, it was analytically found that if variation in height of the consolidating layer was neglected, the time of completion of consolidation due to drainage comprised 50 years; when the variation was taken into consideration, the time was reduced almost fivefold to 10.6 years with the final height of the layer comprising 5.67 m.

4. Creep of Saturated Bases

When a local load is applied on a saturated base, it produces a compound and non-homogeneous state of stress and strain components at each point of the soil mass. If the soil skeleton is capable of creep, then such a base will deform with time, irrespective of the consolidation process. This will result in distribution and redistribution of the stresses between the skeleton and pore water.

Creep of the saturated base determines the initial stage of the consolidation process and is an important step in the solution to problems of consolidation. The stress-strain state of a saturated base under locally applied load will be analysed below with the assumption that the soil skeleton is viscoelastic and the pore water is compressible. The parameters K_{red} and μ_{red} given by the formulae (7.16) and (7.17) respectively can be employed for evaluating the stipulated instantaneous stress-strain state of the soil base. To evaluate the stress-strain state during the period when the soil is experiencing creep, one would, generally speaking, require similar parameters, except that they ought to be time-dependent, i.e., $K_{red}(t)$ and $\mu_{red}(t)$. As regards shearing, it does not depend on pore pressure; therefore, $G_{red}(t) = G_{sk}(t)$.

Let us examine the stress-strain state of an elementary volume of the soil in the absence of drainage. Suppose that at a given instance of time the sum of stresses $\sigma_v = \sigma_x + \sigma_y + \sigma_z$ is acting on the saturated base and the soil skeleton is experiencing creep under volumetric deformation, i.e.,

$$\varepsilon_{sk}(t) = \frac{\sigma'(t)}{3K_m} - \int_{\tau_1}^{t} \frac{\sigma'(\tau)}{3} K(t,\tau) d\tau, \qquad \ldots (7.43)$$

where

$$K(t,\tau) = \frac{\partial}{\partial \tau} \left[\frac{1}{K_m} + \frac{1}{K_l}(1 - e^{-\delta_v(t-\tau)}) \right];$$

and $\sigma'(t) = \sigma - H_w(t)$ is time-dependent effective stress.

Let it be assumed that pore water is linearly compressible and has a modulus of volumetric compressibility $K_w = \text{const}$.

The following condition of compatibility of strain will be applied for determination of $u_w(t)$:

$$\varepsilon_{sk}(t) = \varepsilon(t) = n\varepsilon_w(t). \qquad \ldots (7.44)$$

Taking into consideration formula (7.43) and the equilibrium condition $\sigma'(t) = \sigma - u_w(t)$, it is found that:

$$\frac{\sigma - u_w(t)}{K_m} + \int_{\tau 1}^{t} \frac{[\sigma - u_w(t)]}{K_l} e^{-\delta_v(t-\tau)} d\tau = n u_w(t)/K_w. \qquad \ldots (7.45)$$

The solution to the above integral equation with respect to $u_w(t)$ is:

$$u_w(t) = u_w(\tau_1) \left[1 - \frac{(\delta_v/K_l)(1 - \beta_0)}{A/K_m} \{1 - \exp[-A \times (t - \tau_1)]\} \right],$$

$$\ldots (7.46)$$

where

$$u_w(\tau_1) = \sigma \frac{K_w}{nK_m + K_w} = \sigma\beta_0;$$

$$A = \delta_v \frac{1/K_m + 1/K_l + n/K_w}{1/K_m + n/K_w}.$$

Upon substituting the value of $u_w(t)$ obtained from the solution of (7.46) in eqn. (7.43) and taking into account that $\sigma'(t) = \sigma - u_w(t)$, it follows that:

$$\varepsilon(t) = \varepsilon_{sk}(t) = \frac{\sigma - u_w(t)}{3K_m} - \int_{\tau_1}^{t} \frac{\sigma - u_w(t)}{3} K(t,\tau)\, d\tau.$$

The integration of the above relation yields the following:

$$\varepsilon(t) = \varepsilon_{sk}(t) = \frac{\sigma - u_w(t)}{3K_m} + \frac{\sigma}{3K_l} \frac{1 - e^{-\delta_v(t-\tau)}}{\delta_v} - u_w(\tau_1)\left\{ \frac{1 - e^{-\delta_v(t-\tau_1)}}{\delta_v} \right.$$
$$\left. - B\left[\frac{1 - e^{-\delta_v(t-\tau)}}{\delta_v} - \frac{1 - e^{-(\delta_v+A)(t-\tau_1)}}{\delta_v + A} \right] \right\}, \qquad \ldots \ (7.47)$$

where

$$B = \frac{(\delta_v/K_l)(1 - \beta_0)}{A/K_m};$$

$$\varepsilon(t) = \frac{\sigma}{3}\left\{ \frac{1}{K_m} - \frac{K_w}{nK_m + K_w}(1 - B[1 - \exp(-At)]) + \frac{1 - e^{-\delta_v}(t-\tau)}{\delta_v K_l} \right.$$
$$\left. - \frac{K_w[1 - e^{-\delta_v(t-\tau_1)}]}{(nK_m + K_w)\delta_v}(1 - B) + B\frac{K_w[1 - e^{-(\delta_v+A)(t-\tau_1)}]}{(nK_m + K_w)(\delta_v + A)} \right\}.$$

It follows from the above that:

$$1/K_{red}(t) = 3\varepsilon(t)/\sigma. \qquad \ldots \ (7.48)$$

It is thus seen that a complex functional relationship exists between the reduced bulk modulus of soil under compression and the rheological parameters of the skeleton and compressibility of water. The complexity of the problem becomes obvious if it is taken into account that:

$$\mu_{red}(t) = \frac{K_{red}(t) - 2G(t)}{2[K_{red}(t) - G(t)]}, \qquad \ldots \ (7.49)$$

However, one thing is clear: for the given parameters of soil skeleton, the laws governing the creep deformation of soil as a whole differ from those governing the creep of soil skeleton.

As it is difficult to determine the reduced parameters of the soil as a whole through the parameters of the skeleton and pore water, it is desirable to simplify the problem. For this, it suffices to test the soil specimens in a closed system and relate the parameters to the sum of the total stresses. In this case, the estimation of the stress-strain state of a saturated base during the initial stage is also rendered simpler and is reduced to consideration of the stress-strain state of a quasi-single phase medium. In the problems examined in Chapter 6, this could be achieved by replacing the modulus of deformation of the skeleton with the reduced modulus of whole soil. For example, formula (6.12) or (6.13) could be employed to determine the settlement of soil surface under the action of a local load after replacing E_0, μ_0, G_0 and K_0 by their corresponding reduced values.

If the soil is fully saturated, then $K_w = \infty$ and the volumetric strain is equal to zero. However, settlement is not equal to zero because, in accordance with formula (6.13),

$$s = s_\gamma = \frac{\omega q b}{4G_0(t)}, \qquad \ldots (7.50)$$

where

$$\frac{1}{G_0(t)} = \frac{1}{G_m} + \frac{1}{G_l} \{- \exp[-\delta'_\gamma(t - \tau)]\}.$$

Knowing the state of stress of the soil, the pore pressure can be determined by the following relation:

$$u_w(t) = 3K_w \varepsilon(t)/n, \qquad \ldots (7.51)$$

where

$$\varepsilon(t) = \frac{\sigma}{3} \left\{ \frac{1}{K'_m} + \frac{1}{K'_l} \left[1 - e^{-\delta'_v(t-\tau)} \right] \right\}, \qquad \ldots (7.52)$$

In these relations K'_m, K'_l and δ'_v represent rheological parameters of soil as a whole; $\sigma = (\sigma_1 + \sigma_2 + \sigma_3)/3$.

It is obvious that the above method of determination of pore pressure is much simpler than the one described earlier.

While dealing with the creep of a saturated base whose skeleton displays viscoplastic flow under shear and elastic bahaviour under volumetric strain, the solution to the problem can be obtained either in effective stresses or in total stresses, depending on the deformability and strength parameters employed. If the parameters of soil as a whole, i.e., the reduced parameters K_m, η_{red}, φ_{red} and c_{red} are employed, then the problem is simplified and is basically reduced to that of the estimation of the stress-strain state of a single-phase medium. Such a problem was discussed in Chapter 6. However, if the conventional parameters of soil skeleton are employed, then one has to examine the stress-strain state of two-phase soil. In this approach the field of total stresses and strains is first determined on the basis of parameters K_{red} and μ_{red} and then the pore pressure is determined at an arbitrary point. Now, assuming that the soil has a low coefficient of permeability and high shearing viscosity, the visoplastic strain under shear can be determined under conditions of absence of drainage. The rate of this strain is governed by the shearing stresses and normal effective stresses produced in the saturated base under the action of local load.

Let us examine this solution for a two-dimensional problem. A local load of intensity q is applied on a strip of width $2b$ at the surface of a saturated base. The following soil properties are known: K_{red}, μ_{red}, G, K_w, φ, c and η. First of all, the total stresses are determined by the following relation:

$$\sigma_{1,2} = \frac{q}{\pi} \left\{ \alpha \left[1 - \frac{\beta_w}{3}(1 + \mu_{red}) \right] \pm \sin \alpha \right\}, \qquad \ldots (7.53)$$

where α is the angle of visibility; $\beta_w = K_w/(nK_{sk} + K_w)$.

The pore pressure function can be determined from the formula:

$$u_w(\alpha) = \frac{\beta_w q}{3\pi}(1 + \mu_{red})\alpha. \qquad \ldots (7.54)$$

The following equation describes the viscoplastic flow:

$$\dot{\gamma}_i = (\tau_i - \sigma \operatorname{tg}\varphi - c)/\eta(\sigma), \qquad \ldots (7.55)$$

where

$$\tau_i = (\sigma_1 - \sigma_3)/\sqrt{3}; \quad \sigma'_v = (\sigma'_1 + \sigma'_2)(1 + \mu_{red})/3.$$

The rate of vertical strain is given by the relation:

$$\dot{\varepsilon}_z = \frac{\sigma_z - \sigma}{2\tau_i}\dot{\gamma}_i = \frac{(\sigma_z - \sigma)(\tau_i - \sigma \operatorname{tg}\varphi - c)}{2\tau_i \eta(\sigma)}, \qquad \ldots (7.56)$$

where

$$\sigma_z = \frac{q}{\pi}\left(\operatorname{arctg}\frac{b - x}{z} + \operatorname{arctg}\frac{b + x}{z}\right) - \frac{2bqz(x^2 - z^2 - a^2)}{\pi[(x^2 + z^2 - a^2) + 4a^2 z^2]}; \qquad \ldots (7.57)$$

$$\sigma = (1 + \mu_{red})(\sigma_x + \sigma_z) = (1 + \mu_{red})\frac{2q}{3\pi} \times \left(\operatorname{arctg}\frac{b - x}{z} + \operatorname{arctg}\frac{b + x}{z}\right). \qquad \ldots (7.58)$$

5. Initial Critical Load on Saturated Base

The initial critical load q on a saturated base should be determined by taking into account the pore pressure produced in the soil. Let us examine the stress-strain state in a linearly deformable soil base under plane strain when a strip load is acting on the soil mass. The total stress is given by the following well-known formula:

$$\sigma_{1,2} = \frac{q}{\pi}(\alpha \pm \sin\alpha). \qquad \ldots (7.59)$$

Applying relations (7.59) and (7.18), the pore pressure is determined from the formula

$$u_w(\tau_1) = \frac{\beta_w q}{3\pi}(1 + \mu_{red})\alpha. \qquad \ldots (7.60)$$

The effective stress can then be obtained from the following expression:

$$\sigma_{1,2} = \frac{q}{\pi}\left\{\alpha\left[1 - \frac{\beta_w}{3}(1 + \mu_{red})\right] \pm \sin\alpha\right\}. \qquad \ldots (7.61)$$

It is obvious that at $\beta_w = 0$ which represents the final stage of consolidation, the above expression coincides with the original formula (7.59) for the stabilised state.

The principal stresses due to weight of soil are determined from the condition of isotropic compression, i.e., $\sigma_1' = \sigma_2' = \gamma'(z + h)$ (where h is the depth at which the local load is applied and γ' is the submerged unit weight of soil). Upon substituting these values in (7.61) and the resultant expression in the equation of limiting equilibrium, it is found that:

$$z = \frac{q - \gamma' h}{\pi \gamma'} \left(\frac{\sin \alpha}{\sin \varphi} - B\alpha \right) - \frac{c}{\gamma'} \, \text{ctg} \, \varphi - h, \qquad \ldots \text{(7.62)}$$

where

$$B = 1 - \beta_w (1 + \mu_{red})/3. \qquad \ldots \text{(7.63)}$$

The above equation describes the boundaries of the region of limiting equilibrium. It differs from the well-known equation for single-phase soils and consequently describes a different curve, inside which the condition of limiting equilibrium is satisfied. The maximum ordinate of this region z_{max} is found by equating $z'(\alpha)$ to zero. The expression for α^* is then found as $\alpha^* = \text{arctg}(B \sin \varphi)$. Upon substituting α^* in formula (7.62), the expression for q^* as a function of z_{max} is found as follows:

$$q^* = \frac{\pi(\gamma' z_{max} + \gamma h + c \cdot \text{ctg} \, \varphi)}{\sqrt{1 - B^2 \sin^2 \varphi} / \sin \varphi - B\alpha^*}. \qquad \ldots \text{(7.64)}$$

In the particular case when $\beta_w = 0$ and $B = 1$, the above expression coincides with the known solution of the corresponding problem without consideration of pore pressure [22].

In the case of a fully saturated soil mass, the value of q^*_{min} can be obtained by substituting $\beta_w = 1$, $B = 0.5$ and $\mu_{red} = 0.5$ in eqn. (7.64). For example, if $z_{max} = 0$, the value of q^* for partially saturated and fully saturated bases at $\varphi = 30°$, $c = 0.1$ MPa, $h = 2$ m and $\gamma = \gamma' = 20$ KN/m^3 is found to be 10.2 and 5.64 MPa respectively. This indicates that it is essential to take pore pressure into account while determining the initial critical load because pore pressure leads to the development of large zones of plastic flow in saturated soils. In order to avoid this problem, it is advisable to provide sand drains under the base of the foundation to ensure quick release of the pore pressure. If it is kept in mind that in the initial stage of consolidation the pore pressure increases with time due to creep of the soil skeleton, then it can be concluded that the risk would be maximum not when the loading on the base is completed, but when the extreme pore pressure is attained.

Hence, when a saturated base is designed for the second limiting state based on the theory of a linearly deformable medium, it is first of all necessary to determine the initial critical load.

6. Consolidation and Creep of Saturated Bases under Local Loading

If the intensity of applied load is greater than the initial critical load q^*, then the saturated base experiences plastic deformation and it becomes difficult to describe the stress-strain state by the concepts of the theory of a linearly deformable medium. In this case, it becomes necessary to consider the mixed elastoplastic problem. However, if the region of plastic flow is small compared to the active zone of the stress-strain state, the stress-strain state of a saturated base can be described by means of the theory of linearly deformable space.

A further complication that arises while describing the stress-strain state of a saturated base in the intermediate stage of soil consolidation under local loading is that the sum of the total principal stresses changes with time in the course of consolidation. This factor may be taken into account by the method of successive approximations, which allows correction of the solution to the problem at each step. The condition of constancy of the sum of total principal stresses serves as the first step. Studies conducted by the author have revealed that sometimes the first step itself yields the final solution because the variation of the sum of total stresses is insignificant. For example, from a comparison of the values of the ratio $(1 + \mu_{red})/(1 + \mu)$ it was found that its variation did not exceed 20% when μ_{red} and μ were varied between 0.5 and 0.25. This indicates that the sum of total stresses varies from an initial value to a final value and the difference between the two is 20%. A variation of this magnitude is acceptable for practical calculations.

A few cases dealing with application of local load on a saturated base are discussed below:

Application of concentration force

The reduced parameters E_{red}, μ_{red} and $G_{red} = G$ should be employed for determining the stress-strain state of a linearly deformable base in the initial state. The following expressions are obtained:

—for three-dimensional problem

$$\sigma(\tau_1) = \frac{P}{3\pi} \frac{z}{R^3}(1 + \mu_{red}); \quad w(\tau_1) = \frac{P}{2\pi R} \times \frac{1 - \mu_{red}}{G}; \quad u_w(\tau_1) = \sigma(\tau_1)\beta_w;$$

$$\ldots (7.65)$$

—for two-dimensional problem

$$\sigma(\tau_1) = \frac{2P}{3\pi}(1 + \mu_{red})\frac{z}{r^2}; \quad w(\tau_1) = \frac{P}{\pi}\frac{1 - \mu_{red}}{G} \times \ln\frac{1}{x} + D; \quad u_w(\tau_1) = \sigma(\tau_1)\beta_w,$$

$$\ldots (7.66)$$

where β_w is the coefficient of initial pore pressure determined from formula (7.18).

The intermediate stress-strain state can be evaluated by solving the following equation with the assumption that the sum of the total stresses remains constant.

$$\frac{\partial u_w}{\partial t} = c_v \nabla^2 u_w. \qquad \ldots (7.67)$$

For the adopted initial conditions [see formulae (7.65)] and the boundary conditions $u_w(0, t) = u_w(\infty, t) = 0$, the following relations are obtained:
—for three-dimensional problem

$$u_w(x, y, z, t) = \frac{P}{3\pi} \frac{z}{z^2 + R^2} \beta_w (1 + \mu_{red})$$

$$\times \left[\exp\left(-\frac{z^2 + n^2}{4c_v t} \right) - \frac{1}{\sqrt{z^2 + R^2}} \Phi\left(\frac{z^2 + n^2}{2c_v t} \right) \right], \ldots (7.68)$$

where $\Phi(x)$ is probability integral.
—for two-dimensional problem.

$$u_w(x, z, t) = \frac{2P}{3\pi} \beta_w (1 + \mu_{red}) \left[\exp\left(-\frac{r^2}{4c_v t} \right) - 1 \right]. \qquad \ldots (7.69)$$

Let us examine the stress-strain state of soil half-space under the effect of a local, uniformly distributed load acting on a strip of width $2b$ (plane strain problem) and a rectangle of area $b \times 2b$ (three-dimensional problem). In such cases, the initial stress-strain state can be determined on the basis of the known solutions from the theory of a linearly deformable medium by using E_{red} and μ_{red}. The following expressions are obtained:
—for two-dimensional problem

$$\left. \begin{array}{l} u_w(x, z, \tau_1) = \frac{2}{3} \beta_w (1 + \mu_{red}) \frac{q}{\pi} \left(\text{arctg} \frac{b - x}{z} + \text{arctg} \frac{b + x}{z} \right); \\[3mm] w(x, 0, \tau_1) = \frac{q}{\pi} \frac{(1 - \mu_{red})}{G} \left(2b + \ln \frac{(x - b)^{x-b}}{(x + b)^{x+b}} \right); \end{array} \right\} \quad \ldots (7.70)$$

—for three-dimensional problem

$$\left. \begin{array}{l} u_w(0, 0, z, \tau_1) = \frac{4q}{3\pi} (1 + \mu_{red}) \, \text{arctg} \frac{n}{m\sqrt{1 + m^2 + n^2}}; \\[3mm] w(x, 0, 0, \tau_1) = \frac{bq}{\pi} \frac{1 - \mu_{red}}{G} \left(\ln \frac{\sqrt{1 + n^2} + n}{\sqrt{1 - n^2} - n} + \ln \frac{\sqrt{1 + n^2} + 1}{\sqrt{1 + n^2} - 1} \right), \end{array} \right\}$$
$$\ldots (7.71)$$

where $n = l/b$ and $m = z/(2b)$.

To determine the stress-strain state in the intermediate stage of consolidation, the consolidation equation (7.21) is solved with initial conditions (7.18) and (7.19) for the two- and three-dimensional problems respectively. A closed form solution of the equation is not possible. The change in distribution of pore pressure with time for the soil column in the middle portion of the area under load can be obtained by assuming that consolidation by drainage occurs in the above-mentioned column only in the vertical direction.

The method of source function on a infinite straight line will be applied to obtain the solution. According to this method [18], if the initial distribution of pore pressure along the height of the soil column and the boundary conditions $u_w(0,t) = u_w(0,0,t) = 0$ are known, then the solution of (7.21) can be written in the form given below:

$$u_w(z,t); = \frac{1}{2\sqrt{c_v t \pi}} \int_0^\infty u_w(\zeta,0) \left\{ \exp\left[-\frac{(z-\zeta)^2}{4c_v t}\right] - \exp\left[-\frac{(z+\zeta)^2}{4c_v t}\right] \right\} d\zeta,$$

$$\dots (7.72)$$

where $u_w(z,0)$ is the initial distribution of pore pressure.

In accordance with expressions (7.70), we get the following relations:
—for two-dimensional problem

$$u_w(z,t) = \frac{2q}{3\pi}\beta_w \frac{1+\mu_{red}}{\sqrt{\pi c_v t}} \int_0^\infty \text{arctg}\,\frac{b}{\zeta} \left\{ \exp\left[-\frac{(z-\zeta)^2}{4c_v t}\right] \right.$$

$$\left. - \exp\left[-\frac{(z+\zeta)^2}{4c_v t}\right] \right\} d\zeta; \qquad \dots (7.73)$$

—for three-dimensional problem

$$u_w(z,t) = \frac{4q}{3\pi} \frac{1+\mu_{red}}{\sqrt{\pi c_v t}} \times \int_0^\infty \text{arctg}\,\frac{l/b}{[\zeta/(2b)]\sqrt{1+\zeta^2/(4b^2)+l^2/b^2}}$$

$$\times \left\{ \exp\left[-\frac{(z-\zeta)^2}{4c_v t}\right] - \exp\left[-\frac{(z+\zeta)^2}{4c_v t}\right] \right\} d\zeta. \quad \dots (7.74)$$

The integration of the above equations is beset with many difficulties; therefore, it is convenient to represent the non-linear function $f(z)$ of initial pore pressure by an equivalent straight line, such that the area of the triangular diagram is equal to that of the curvilinear pore pressure diagram. In this case, the following relations are obtained:
—for two-dimensional problem

$$h_{red} = \frac{4}{\pi} \int_0^\infty \text{arctg}\,\frac{b}{\zeta}\,d\zeta; \qquad \dots (7.75)$$

—for three-dimensional problem

$$h_{red} = \frac{4}{\pi} \int_0^\infty \text{arctg}\,\frac{n}{m}\,\frac{1}{\sqrt{1+m^2+n^2}}\,d\zeta, \qquad \dots (7.76)$$

where $n = l/b$ and $m = \zeta/(2b)$.

The initial distribution of pore pressure will be described by the following formulae:

—for two-dimensional problem

$$u_w(z, \tau_1) = \frac{2}{3}\beta_w \frac{(1 + \mu_{red})}{\pi} q\left(1 - \frac{z}{h}\right) = A_{pl}q\left(1 - \frac{z}{h_{pl}}\right); \qquad \ldots \text{(7.77)}$$

—for three-dimensional problem,

$$u_w(z, \tau_1) = \frac{4q}{3\pi}(1 + \mu_{red})\left(1 - \frac{z}{h_{pl}}\right) = A_{red}q\left(1 - \frac{z}{h_{red}}\right). \qquad \ldots \text{(7.78)}$$

Upon substituting the above values in formula (7.72) and integrating, the following expression is obtained for the two-dimensional problem:

$$u_w(z, t) = A_{pl}q\left[\Phi\left(\frac{z}{2\sqrt{c_v t}}\right) - \frac{1}{2}\left(1 - \frac{z}{h_{pl}}\right) \times \Phi\left(\frac{z - h_{pl}}{2\sqrt{c_v t}}\right)\right.$$

$$- \frac{1}{2}\left(1 + \frac{z}{h_{pl}}\right)\Phi\left(\frac{z + h_{pl}}{2\sqrt{c_v t}}\right) + \frac{\sqrt{c_v t}}{h_{pl}\sqrt{\pi}}$$

$$\left. \times \left\{\exp\left[-\frac{(z - h_{pl})^2}{4c_v t}\right] - \exp\left(-\frac{(z + h_{pl})^2}{4c_v t}\right)\right\}\right] \qquad \ldots \text{(7.79)}$$

Wherefrom, it follows that:

$$u_w(z, 0) = A_{pl}q(1 - z/h_{pl}); \quad u_w(z, 0, 0) = 0;$$

$$U(t) = 1 - A_{pl}\left\{\left(3 + \frac{2c_v t}{h_{pl}^2}\right)\Phi\left(\frac{h_{pl}}{2\sqrt{c_v t}}\right) - \left(2 - \frac{c_v t}{h_{pl}^2}\right) \times \Phi\left(\frac{h_{pl}}{\sqrt{c_v t}}\right)\right.$$

$$\left. + \frac{2\sqrt{c_v t}}{h_{pl}\sqrt{\pi}}\left[3\exp\left(-\frac{h_{pl}^2}{4c_v t}\right) - \exp\left(-\frac{h_{pl}^2}{c_v t}\right) - 2\right]\right\}. \qquad \ldots \text{(7.80)}$$

Similar expressions can be obtained for the three-dimensional problem by substituting A_{sp} in place of A_{pl}.

The ratio of the areas of the $\sigma_z(t)$ and σ_z diagrams represents the degree of consolidation, i.e.,

$$U(t) = \frac{2}{qh}\int_0^h \left[q\left(1 - \frac{z}{h}\right) - u_w(z, t)\right]dz. \qquad \ldots \text{(7.81)}$$

It can be seen from Fig. 7.4 that curves 1 and 3 are closer to each other than to curve 2. Hence, the proposed method with boundary conditions $u_w(t, \infty) = 0$

148

and $u_w(t,0) = 0$ is closer to the one-dimensional problem of equivalent layer with two-way drainage. From calculations based on the proposed method and the method of equivalent layer with one-way drainage, it is found that for a homogeneous soil base, the duration of the consolidation process by the proposed method is fourfold less. It may be mentioned that the boundary conditions for a homogeneous soil base adopted in the proposed analytical scheme accord better with the actual drainage conditions.

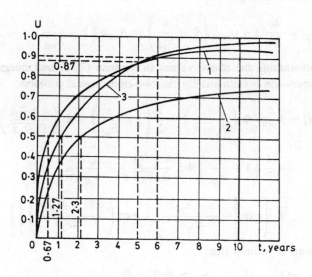

Fig. 7.4. Variation of the degree of consolidation with time.

1—curve based on the proposed method; 2 and 3—curves based on the method of equivalent layer with one-way and two-way drainage respectively.

The variation of settlement with time can be evaluated using $u(t)$ by the following formulae:

$$s(t) = s_\gamma + s_v u(t), \qquad \ldots (7.82)$$

where

$$s_\gamma = \sum_{j=1}^{j=n} \frac{\sigma_{zj} - \sigma_j}{2G_j} \Delta h_j; \quad s_v = \sum_{j=1}^{j=n} \frac{\sigma_j}{k_j} \Delta h_j. \qquad \ldots (7.83)$$

The following equation can be employed for determining the settlement of the surface of a base whose consolidation has not stabilised:

$$\varepsilon_z(t) = \frac{\sigma'_z(t) - \sigma'(t)}{2G} + \frac{\sigma'(t)}{K}. \qquad \ldots (7.84)$$

As $\sigma'_z(t) = \sigma_z - u_w(t)$ and $\sigma'(t) = \sigma - u_w(t)$, it is found that:

$$\varepsilon_z(t) = \frac{\sigma_z - \sigma}{2G} + \frac{\sigma - u_w(t)}{K}. \qquad \ldots (7.85)$$

It is evident from the above that in the intermediate stage of consolidation of a linearly deformable base of saturated clayey soil, the settlement increases with time only due to volumetric deformation of soil. In this case, the shearing deformation is independent of the variation of pore pressure with time. If the shearing modulus of soil is a function only of the sum of the principal effective stresses, then the shearing deformation will depend on the rate of dissipation of pore pressure. However, during consolidation the soil is compressed and hardened; therefore, the sum of the effective principal stresses increases. Consequently, the initial period of consolidation is most favourable for the development of shearing deformation.

It is much more difficult to determine the variation of settlement with time if the creep of soil skeleton is taken into account. In the first approximation, the creep due to volumetric deformation during consolidation by drainage may be ignored and the time-dependent settlement may be expressed as

$$s(t) = s_\gamma(t) + s'_v u(t) + s''_v, \qquad \ldots (7.86)$$

where

$$\left. \begin{array}{l} s_\gamma(t) = \displaystyle\sum_{j=1}^{j=n} \frac{\sigma_{zj} - \sigma_j}{2G_j(t)}; \\[3mm] s'_v = \displaystyle\sum_{j=1}^{j=n} \frac{\sigma_j}{K_j}; \quad s''_v = \displaystyle\sum_{j=1}^{j=n} \frac{\sigma_j}{K_j(t)}. \end{array} \right\} \qquad \ldots (7.87)$$

Application of local load on soil mass of finite thickness and width

The problem is determination of the stress-strain state of a soil mass of limited dimensions under the action of a local load (Fig. 7.5). Let us first examine how the initial distribution of pore pressure is determined. For this it would be necessary to solve the problem of distribution of mean total stress in the soil mass. It is known, that in two-dimensional problems the sum of the principal stresses should satisfy the Laplace equation, i.e.,

$$\frac{\partial^2 \sigma}{\partial x^2} + \frac{\partial^2 \sigma}{\partial z^2} = 0, \qquad \ldots (7.88)$$

where $\sigma = (\sigma_x + \sigma_z)(1 + \mu)2/3$.

The boundary conditions for this problem are as follows:

$$\sigma(x, 0) = q(1 + \mu_{\text{red}})2/3; \quad 0 \le x \le b;$$

$$\frac{\partial \sigma}{\partial x}(0, z) = 0; \quad \frac{\partial \sigma}{\partial x}(l, z) = 0.$$

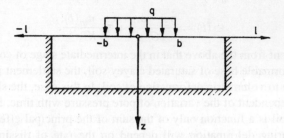

Fig. 7.5. Analytical scheme of consolidation of a soil mass of finite thickness and width.

For the above boundary conditions, the final expression of function $\sigma(x, z)$ is obtained in the form:

$$\sigma(x, z) = \frac{2q}{3}(1 + \mu_{red})\left(\frac{c}{l} + \frac{2}{\pi}\sum_{n=1}^{\infty}\frac{1}{n} \times A_n / sh\frac{n\pi 2h}{l}\right), \quad \dots (7.89)$$

where

$$A_n = \left[sh\frac{n\pi z}{l} + sh\frac{n\pi(2h - z)}{l}\right]\sin\frac{n\pi c}{l} \times \cos\frac{n\pi x}{l}.$$

The stress components can be determined with the help of the well-known expressions [21]:

$$\tau_{xz} = -x\frac{\partial\sigma}{\partial z}; \quad \sigma_x = \sigma + x\frac{\partial\sigma}{\partial x}; \quad \sigma_z = \sigma - x\frac{\partial\sigma}{\partial x}. \quad \dots (7.90)$$

The initial distribution of pore pressure in the region under consideration will be given by the expression:

$$u_w(x, z, 0) = \sigma(x, z)\beta_w, \quad \dots (7.91)$$

where $\sigma(x, z)$ is determined with the help of function (7.89) and β_w from formula (7.18).

Assuming further that the sum of principal stresses remains constant during consolidation (according to V.A.Florin), the equation of consolidation for the two-dimensional problem is obtained in the form (7.79). If the initial distribution is described by (7.91) and the boundary conditions are $u_w = 0$ at $z = 0$, $z = h$, $x = \pm l$, the solution is obtained as follows:

$$u_w(x, z, t) = \sum_{i,j=1}^{\infty} A_{ij}\exp(-c_v\pi^2)\left(\frac{i^2}{l^2} + \frac{j^2}{h^2}\right)\sin\frac{i\pi x}{l}\sin\frac{j\pi z}{h}, \quad \dots (7.92)$$

where

$$A_{ij} = \frac{4}{lh} \int\limits_0^l \int\limits_0^h u_w(x, y, 0) \sin \frac{i\pi x}{l} \sin \frac{j\pi z}{h} dx dz. \qquad \ldots (7.93)$$

Thus the problem formulated at the beginning has been completely solved.

8

Short-term and Long-term Stability
of Slopes

1. General Principles

The problems of short-term and long-term stability of slopes and embankments are inviting the increasing attention of engineers because of recent interest in hilly regions that are rich in mineral and hydropower sources. The enhanced interest in these problems can also be attributed to activities related to the construction of transportation highways, retaining walls, pipelines, tunnels etc. The solutions to these problems require refinement of the existing methods of quantitative estimation of the short-term and long-term stability of slopes and embankments. It is also necessary to review the dynamics of landslide processes by taking into account the interaction between the soil mass of the sliding slope and the engineering structure. This helps in preventing the dangerous consequences of land slides and in developing economically feasible measures that provide stability to the slopes and ensure normal service conditions for the structure.

A number of solutions and methods of qualitative and quantitative estimation of the stability of slopes and embankments are presently available. However, quantitative estimation of displacements occurring during landslides has not received sufficient attention, although this factor is of paramount importance in ensuring normal working conditions for structures built on slopes.

In most cases, the slopes have adequate long-term and short-term stability, i.e., displacements of catastrophic magnitudes do not occur at a given time and during the service life of a structure. However, the soil mass of a slope can experience slow deformation in space and time due to the stresses and strains produced by shearing stresses and other factors that contribute to sliding. These deformations produce movements on the soil surface which are recorded by bench-marks on the surface. Such displacements are insignificant in magnitude, ranging from a few millimeters to a few centimetres in a month or year. However, for many structures that are sensitive to non-uniform deformation, these displacements may exceed the limits permissible for normal service of the structure. In view of the foregoing, it is necessary to design slopes and embankments on the basis of the first as well as the second group of limit states. In the first case, it is required to determine the probability of the occurrence of instability and catastrophic displacement of the soil mass of the slope, the volume of simultaneously sliding mass and the time of onset of catastrophic displacement (landslide). In the second case, it becomes essential to establish the stress-strain state of the slope before and after the engineering activity, the magnitude and velocity of displacement at individual points of the slope and the possibility of transition from the prolonged creep phase to a catastrophically progressing phase due to large build-up of strain in the slope and reduction of strength with time.

Generally speaking, the classification of slopes into stable and unstable is based on the time factor and developing displacements. A slope which appears to be stable for a short duration may become unstable with the passage of time due to large deformations and displacements of soil because during this process the strength of soil drops to the residual strength level. In view of this, the design of slopes should be based upon a unified approach with simultaneous consideration of deformability and stability since the processes of deformation and breakdown of the soil comprising the slope are organically linked to each other.

It is extremely difficult to suggest a single estimate of the sliding process covering both the deformation phase and the phase of catastrophic displacement. This is because rheological testing is not easily done of soils comprising the slope nor determination of its stress-strain state, taking into account the important factors that produce sliding. The problem is further compounded by the fact that natural slopes have a stress-strain state which has evolved over a period of time. Hence, while analysing such slopes one has to deal not with the initial stage of the sliding process as in earth structures and pit walls, but with the intermediate stage.

In short, quantitative estimation of slippage processes is presently in the nascent stage of development and future success in this field will depend on the solutions to the following two problems: reliable determination of the rheological parameters of soils comprising the slope and reliable estimation of the stress-strain state of slope, taking into account its formation history and the factors that produce sliding.

2. Fundamentals of Analysis of Slopes according to First and Second Limit States

On analysis of slopes according to the first and second limit states can presently be carried out on the basis of certain criteria that define the limit state in terms of strength or stability (first limit state) or in terms of deformation (second limit state).

Criteria of analysis of slopes according to first limit state

The overall stability of slopes can be quantitatively estimated by means of the factor of safety which represents the ratio of resisting forces (moments) M_k that are stabilising a particular soil mass to the forces (moments) M_{ov} that tend to overturn it, assuming that the given volume of soil slides as a rigid mass over a cylindrical surface, i.e.,

$$\eta_{stab} = M_k / M_{ov}. \qquad \qquad \dots (8.1)$$

Similar coefficients are employed for more complex forms of displacement, e.g., displacement over a sliding surface in the form of a straight or a broken line. In such cases, the following ratio between the resisting and disturbing forces is considered:

$$\eta_{stab} = T_k / T_{ov}. \qquad \qquad \dots (8.2)$$

With the help of the above coefficients it is possible to assess the stability of a slope in the first approximation, provided the parameters defining soil strength are known. If the parameters of long-term strength of soil are known, then the same criteria can be applied for assessment of the short-term and long-term stability of slopes over a given period of time.

The ratio of the stress and strain at an arbitrary point of the soil mass to the limiting value, i.e., the margin of safety at each point of the soil mass, can be adopted as the criterion for evaluating the condition of a slope. The factor of safety for strength can be used in determining the safety factor in terms of overall stability of the soil mass along any arbitrary surface of sliding. For example, the margin of safety at any point of the soil mass can be determined using the stresses by the following formula for factor of safety with respect to strength:

$$\eta_{str} = \frac{\tau_\nu^*}{\tau_\nu} = \frac{2\tau_{max} \cos \varphi}{c + \mathrm{tg}\,\varphi(\sigma_1 + \sigma_2 - 2\tau_{max} \sin \varphi)}, \qquad \dots (8.3)$$

where c is cohesion and φ is the angle of internal friction.

The above expression is obtained by considering the Mohr circles for the given region in the limit state and prior to the onset of a limit state.

Isolines of the factor of safety of soil in the mass under consideration, which give a graphic representation of the state of soil mass at any point can be plotted by means of formula (8.3). The factor of safety of soil can also be determined on the basis of other strength theories. One such theory is the strength theory that correlates the limit state with limiting angular strain γ^* or the limiting rate of angular strain $\dot{\gamma}^*$ when the shearing strength of soil becomes maximum or minimum, i.e., it attains the peak value or drops to the residual strength level. Then the ratio of limiting values γ^* and $\dot{\gamma}^*$ to their values prior to the limit state γ and $\dot{\gamma}$ can serve as the criterion for evaluation of the factor of safety with respect to strength:

$$\eta_{str} = \gamma^*/\gamma; \quad \eta_{str} = \dot{\gamma}^*/\dot{\gamma}. \qquad \ldots (8.4)$$

If $\gamma^* = \gamma_{max}^*$, then the possibility of failure of the slope at the given moment is evaluated. But if $\gamma^* = \gamma_{res}^*$, then the condition of the slope over a given duration is assessed, keeping in mind that the angular deformation at an arbitrary point of the soil mass $\gamma(t, x, y, z)$ can be estimated for the above duration. The limiting values of shearing strain γ^* and rate of shearing strain $\dot{\gamma}^*$ for the given type of soil can be determined on the basis of the results of laboratory experiments conducted according to the kinematic scheme. As regards the values of shearing strain prior to the onset of limit state, these can be estimated from the stress-strain state of slope, taking into account the rheological shearing of the soil mass.

It is thus seen that the last two criteria for evaluating the stability of a slope require estimation of its stress-strain state, which is not possible without the help of numerical methods.

Criteria of analysis of slopes according to second limit state

When a sliding slope interacts with an engineering structure (pipeline, road, bridge, embankment wall etc.), it becomes necessary to evaluate the stress-strain state of the slope to ensure strength, durability and normal service conditions for the structure under the possible creep deformation of the slope prior to the attainment of strength limit. Obviously, the criterion adopted in these cases should be the limiting displacement or the difference between displacements at individual points of the structure interacting with the sliding slope.

$$\left.\begin{array}{c} \text{at} \quad u(t) \leq u^* \\ \Delta u(t) \leq \Delta u^*; \\ \text{at} \quad v(t) \leq v^* \\ \Delta v(t) \leq \Delta v^*, \end{array}\right\} \qquad \ldots (8.5)$$

where u^* and v^* are limiting displacement of structure in the horizontal and vertical direction respectively; $u(t)$ and $v(t)$ are time-dependent design values of displacement of structure in the horizontal and vertical direction respectively.

It is obvious that for calculating $u(t)$ and $v(t)$, it is necessary to evaluate the stress-strain state of the sliding mass, taking into account the factors that produce sliding, the rheological properties of soil and their interaction with the structure. From the known values of soil deformation, the displacement of a point on the surface of a sliding slope can be determined with respect to a stipulated stationary point at a given depth by integration or summation of the elementary deformations:

$$u(t) = \int_{x_0}^{x} \varepsilon_x(t)dx; \quad v(t) = \int_{y_0}^{y} \varepsilon_y dy. \qquad \ldots (8.6)$$

In some cases (for example, indefinite creep of slopes). the rates of deformation of soil and displacement can be determined by applying the principle of incompressible viscous body, while the strain rate can be determined by means of the viscous flow equation:

$$\left. \begin{array}{l} \varepsilon = \dfrac{\partial u}{\partial x} + \dfrac{\partial v}{\partial y} = 0; \quad \sigma_x - \sigma = 2\eta\varepsilon_x; \\ \sigma_y - \sigma = 2\eta\varepsilon_y; \quad \tau_{xy} = \eta\gamma_{xy}, \end{array} \right\} \qquad \ldots (8.7)$$

where η is viscosity.

The displacement over a given period of time can also be determined from the flow velocity:

$$u(t) = \frac{t}{4\eta} \int_{x_0}^{x} (\sigma_x - \sigma_y)dx; \quad v(t) = \frac{t}{4\eta} \int_{y_0}^{y} (\sigma_y - \sigma_x)dy. \qquad \ldots (8.8)$$

To determine the displacements and their rates in the case of slopes of pits and trenches, it is necessary to apply more advanced rheological models that take into account the unsteady nature of the growth of creep in soils comprising the slope, particularly in the initial stage of formation of the slope.

The methods described above for analysis of slopes according to the first and second limit states may be applied in engineering practice to develop measures aimed at stabilisation of sliding processes and ensuring normal working conditions for structures built on slopes.

3. Analysis of Stability of Slopes and Embankments

While analysing the stability of slopes, the greatest difficulty is encountered in predicting the onset of catastrophic displacement because this involves evaluation of the stress-strain state of sliding mass taking into account the rheological behaviour of soil. However, the solution to this problem can be simplified if it is assumed that the shearing strength of soil is mobilised first of all along the future slip surface, which is invariably cylindrical in shape.

Several methods have been developed for analysis of slopes according to the first limit state. These include approximate methods as well as exact methods based on analysis of the stress-strain state.

Method of cylindrical slip surface (MCSS)

This is the simplest method for evaluating the factor of safety of slopes; hence it is widely used in engineering practice. A comparison of the analytical results obtained by this method with those of other more accurate methods [17, 21] based on evaluation of the stress-strain state revealed that there is not much difference between the two. However, as MCSS is relatively much simpler and requires considerably less computational efforts, it is irreplaceable for preliminary calculations.

The main drawback of MCSS is that it does not satisfy all the three equilibrium equations, i.e., it does not take into account the mutual effect of individual slices and their deformability during failure of the slope. However, this shortcoming is fully compensated by the fact that by the MCSS method it is possible to determine the factor of safety for stability of inhomogeneous soil mass of an arbitrary surface profile, taking into account the effect of seepage and seismic and other forces.

The modern methods of stability analysis are based on an evaluation of the two-dimensional problem of soil mechanics, which does not always conform to the actual field conditions. In most cases, the sliding mass has a spherical sliding surface. As noted by many researchers, if this aspect is taken into consideration, it can significantly increase the factor of safety. When the stability is analysed within the framework of a two-dimensional problem, the factor of safety is lower, thereby resulting in a greater margin of stability.

The method essentially involves determination of the ratio of the moment of forces holding the soil mass to the moment of forces tending to overturn it about the centre of the arbitrarily selected cylindrical sliding surface (Fig. 8.1). In this case, the factor of safety is determined as follows:

$$\eta_{\text{stab}} = M_k/M_{ov} = \sum_{i=1}^{i=n} (N_i f_i + c_i L_i) / \sum_{i=1}^{i=n} T_i, \qquad \dots (8.9)$$

where $N_i = Q_i \cos \alpha_i$; $T_i = Q_i \sin \alpha_i$; $f_i = \text{tg} \varphi_i$ (here α_i is the angle which the base of the slope makes with the horizontal, Q_i is the mass of section); L_i is the length of sliding path in the ith section

As the factor of safety is determined for an arbitrary sliding surface, it is desirable to examine a few surfaces with different positions of the centre of rotation and then select the most critical sliding surface. This procedure can be easily implemented by using a standard computer program based on formula (8.9).

In the method described above, the inertial (seismic) forces can be taken into consideration by applying the principle of quasi-static equivalent force, i.e., $N_{\text{seis}} = N_{st} k_{\text{seis}}$ (where k_{seis} is the seismicity coefficient determined as the ratio of the gravitational acceleration to the acceleration of the seismic force; N_{st}, N_{seis} are the static and seismic forces respectively). While implementing

158

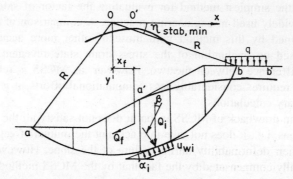

Fig. 8.1. Analytical scheme for determining the factor of safety by the method of cylindrical slip surface.

the above procedure, the most unfavourable direction of action of the seismic force (from the standpoint of stability) on the given slope should be considered.

In general, when the seismic force is acting at an angle β to the vertical and is directed towards the slope, it is found that:

$$\eta_{stab} = \frac{\displaystyle\sum_{i=1}^{i=n} \{Q_i[\cos\alpha_i + k_{seis}\cos(\alpha_i + \beta)]f_i + c_i L_i\}}{\displaystyle\sum_{i=1}^{i=n} Q_i[\sin\alpha_i + k_{seis}\sin(\alpha_i + \beta)]} \qquad \ldots (8.10)$$

While analysing the stability of permeable slopes, it is necessary to take the seepage forces into account because they reduce the margin of stability. The method for determination of seepage forces has been described in [17].

The effect of pore pressure on the stability of slopes should be taken into consideration when the consolidation and seepage processes are not steady due to changes in the stress-strain state and pressure in the saturated water-table level. In the case of slopes of earth structures with light seepage (dams, banks, canals etc.), the pore pressure during the initial (construction) period should be determined under absence of drainage. During analysis of stability, the pore pressure may be taken into account by applying the principle of effective stresses, using the strength parameters determined from tests conducted according to the consolidated-drained test scheme. In such cases, analysis of the stability of slopes is carried out in terms of total stresses; the pore pressure is taken into account indirectly, using the strength parameters determined from tests conducted according to unconsolidated-undrained scheme. In calculating the factor of safety, consideration of pore pressure involves its subtraction from the total normal stress acting on the sliding surface, i.e.,

$$\eta_{\text{stab}} = \sum_{i=1}^{i=n}[(N_i - u_{w_i}L_i)f_i + c_iL_i]\sum_{i=1}^{i=n}T_i, \qquad \ldots (8.11)$$

where u_{w_i} is pore pressure in the ith slice.

It is evident that pore pressure reduces the factor of safety of the slope.

Sometimes the sliding surface may come into contact with water-bearing strata of the soil mass having varying pressure at points of contact with the slip surface. In such cases, while carrying out the stability analysis, the pressure in the water levels is substituted for pore pressure in formula (8.11).

Obviously, there is considerable scope for application of the method of cylindrical slip surface in the analysis of the stability of slopes since in this method one can take into account such factors as are impossible to consider in any other method.

Method of limiting state of stress (MLSS)

This method of analysis of the stability of slopes has found wide application in engineering practice, due mainly to the efforts of Soviet scientists. The method is based on the theory of limiting equilibrium of soil medium. Unlike the theories dealing with the stresses and strains before the attainment of a limit state, the theory of limiting equilibrium considers the equations of equilibrium along with the condition of limiting equilibrium. Instead of four arbitrary functions however, just two are employed to satisfy the boundary conditions. This implies that in solving the problems of stability of slopes by the theory of limiting equilibrium, it is not always possible to satisfy all the boundary conditions. Hence the solution obtained is sometimes discontinuous or mixed, i.e., it is based on the assumption that prelimit state regions also exist. The last case better reflects the actual behaviour, as it is difficult to imagine the limiting equilibrium state appearing simultaneously at all points of the soil mass under consideration. However, in the absence of efficient methods for solving a mixed problem, one is constrained to apply the theory of limiting equilibrium to the whole region under scrutiny.

In analysing the stability of slopes, the theory of limiting equilibrium is applied within the framework of a two-dimensional problem for which the system of equations is established. In this case, there are two equations of equilibrium with three unknowns:

$$\frac{\partial \sigma_x}{\partial x} + \frac{\partial \tau_{xy}}{\partial y} = \gamma; \qquad \frac{\partial \sigma_y}{\partial y} + \frac{\partial \tau_{xy}}{\partial x} = 0, \qquad \ldots (8.12)$$

and another of limiting equilibrium which closes the system

$$(\sigma_x - \sigma_y)^2 + 4\tau_{xy}^2 = (\sigma_x + \sigma_y + 2c + \text{ctg}\,\varphi)^2 \sin^2 \varphi, \qquad \ldots (8.13)$$

V.V. Sokolovskii reduced the solution to the problem to a system of two second-order equations. On the basis of the solution obtained by him, a number of analytical methods have been recommended for determination of maximum pressure on the horizontal soil surface (at which the slope of a given profile remains in equilibrium) and the forms of equistable slopes of limiting gradient.

Method of limiting stress-strain state (MLSSS)

This method was widely applied during the last few decades for analysis of slopes according to the firsts limiting state [19]. Unlike the preceding method, in the MLSSS it is assumed that the soil mass under consideration is, on the whole, in a prelimit state, although the limit state is prevalent at individual points or local (closed) sections of the soil mass. The author believes this assumption accords better with the stress-strain state of slopes having a factor of safety greater than unity. In such cases, an analysis of the stress-strain state helps in identifying the zones of possible plastic flow and in determining the potential sliding surface, tasks difficult to achieve by the other methods. This is particularly important if an elasto-viscoplastic hardening porous medium is adopted as the rheological model of soil. However, in this case, the evaluation of stress-strain state becomes much more complicated and the solution requires a much greater volume of computer calculations. Yet, this method enables determination not only of the zones of limit state, but also the displacement and rate of displacement at individual points of the soil mass.

Thus, the MLSSS enables evaluation of the stress-strain state of slope and its limit state on the one hand, and estimation of slide displacements and their rates, on the other. It may be mentioned though that such a unified analysis is not always possible. It requires that the rheological model of soil should include all the stages of deformation (transient, steady, damping or progressing) or soil, a task associated with tremendous, sometimes insurmountable, difficulties.

For the simplest case of linearly deformable medium, the stress-strain state of a slope of inhomogeneous composition is also evaluated by numerical method on a computer. The zones of limit stress-strain state are determined from condition (8.3) or (8.4).

For a homogeneous and isotropic soil mass with a curvilinear boundary, the stress-strain state is evaluated analytically by applying the Kolosov-Muskhelishvili method of complex potentials [17, 19]. Numerous evaluations of the stress-strain state of homogeneous soil mass with a curvilinear boundary have been obtained by computer calculations applying the method described above [17, 19]. An analysis of the calculations revealed that the assumption regarding the cylindrical shape of the sliding surface of the sliding mass is justified in many cases. Besides, the factors of safety for the cylindrical sliding surface, computed on the basis of stresses determined by this method, were found to be close to those calculated by the usual method.

In the given method, it is necessary to take pore pressure into account while evaluating the stress-strain state of saturated soils during the initial period of their intensive loading, and also in the case of undercutting and overloading of slopes and rapid erection of embankments on a saturated base. The easient way to do this is by means of reduced deformability parameters E_{red}, μ_{red} and reduced strength parameters φ_{red}, c_{red} which allow the pore pressure to be taken into account implicitly. To take pore pressure into account in an explicit form, it is necessary to change over from total stresses to effective stresses and replace the reduced strength parameters by effective strength parameters φ and c.

To evaluate the long-term stability of slopes, it is necessary to determine the anticipated changes in their stress-strain state in space and time due to variations in the temperature and moisture content fields and interaction with the structure and the surrounding geological medium. In the first approximation, the long-term stability of slopes can be evaluated by accounting for the time-dependent variation of the strength properties of the soils in the mass.

In such cases, the factor of safety for stability can serve as the criterion for estimation of long-term stability. It may be pointed out that the factor of safety for stability changes with time as the soil mass experiences the complex process of distribution and redistribution of stresses due to rheological phenomena and consolidation and also variations in the surrounding medium. It is obvious that it would be extremely difficult to consider all the aforementioned features while evaluating the long-term stability; hence, only a few decisive factors are taken into account. It may be further noted that evaluation of long-term stability on the basis of limiting state of stress and limiting state of strain involves the solution of fundamentally different problems of soil mechanics.

The long-term stability of slopes based on limiting state of stress should be considered, first of all, in those cases wherein the stress-strain state of soil is seriously disturbed due to undercutting, overloading and reworking of the soil etc., and also due to significant changes in the hydrological conditions. Here we shall discuss some of the established methods of evaluation of long-term stability in which it is possible to consider the effect of a number of factors.

By the method of cylindrical slip surface, it is possible to evaluate the long-term stability of slopes with consideration of the time-dependent strength properties of soil, hydrological conditions and moisture content.

On the basis of formula (8.9), the factor of safety for long-term stability can be represented in the form:

$$\eta_{stab}(t) = \sum_{i=1}^{i=n} \left\{ [N_i - u_{w_i}(t)L_i] f_i(t) + c_i(t)L_i \right\} / \sum_{i=1}^{i=n} T_i. \qquad \ldots (8.14)$$

It is evident that the above coefficient is time-dependent, it also depends on the change in soil strength with time, i.e., $\varphi(t)$ and $c(t)$ as well as the change in pore pressure with time $u_w(t)$.

The method of limiting state of stress may be applied for evaluation of the long-term strength of slopes and in solving the following problems:
— Determination of the diagram of limiting load on the horizontal surface of a straight slope so as to ensure its equilibrium over a given period of time.
— Determination of the profile of an equistable slope at the given instant of time whenever the soil displays both friction and cohesion.
— Determination of load intensity on the horizontal surface of slope.
— Determination of the rate of displacement at an arbitrary point of the slope which is in a state of viscoplastic flow.

The simplest way to determine the diagram of limiting load on the horizontal surface of the slope is based on the solution given by V.V. Sokolovskii [22], provided the variation of strength parameters of soil with time is known:

$$q(y, t) = \overline{\sigma}_z c(t) + q_c(t); \quad y(t) = \overline{y}[c(t)/\gamma], \qquad \ldots (8.15)$$

where $\overline{\sigma}_z$ is non-dimensional limiting load, determined from the table given in [22] and $\overline{y}$ is the relative co-ordinate, determined from the same Table; $q_c(t) = c(t) + \text{ctg}\,\varphi(t)$.

By assigning various values of time $t_1, t_2, \ldots, t_n$ and calculating the corresponding values of c and φ, it is possible to determine the diagram of limiting load on the horizontal surface of a straight slope for a given period of time, after which the slope may fail.

The profile of equistable slope is also determined on the basis of the solution proposed by V.V. Sokolovskii [22]. For instance, for the case wherein the soil displays both cohesion and friction, the profile of the equistable slope is described by the following equation:

$$x(t) = \overline{x}[c(t)/\gamma]; \quad y(t) = \overline{y}[c(t)/\gamma], \qquad \ldots (8.16)$$

where $\overline{x}$ and $\overline{y}$ are non-dimensional co-ordinates that depend upon φ and c [22].

The intensity of limiting load on the horizontal surface of an equistable slope may be determined by the formula:

$$q(t) = \frac{2c(t) \cos \varphi(t)}{[1 - \sin \varphi(t)]}$$

or,

$$q_c = \frac{2c_l \cos \varphi_l}{(1 - \sin \varphi_l)},$$

where c_l and φ_l are parameters that describe the limit of long-term strength of soil.

At $\varphi = 0$, it is found that,

$$q_l(t) = 2c(t); \quad q_l = 2c_l.$$

To solve the last of the problems listed above, it is necessary to evaluate the state of stress of the soil mass which is in a state of limiting equilibrium, and to determine the field of rate of displacement based on viscous resistance to shear. The solution to this problem is extremely difficult to obtain and depends on the shape and dimensions of the soil mass, homogeneity of the soil mass etc.

As shown by V.V. Sokolovskii and M.V. Malyshev [12], the stress-strain state of soil mass in the state of limiting equilibrium can be evaluated using a stress function which exactly satisfies the equations of equilibrium but only approximately satisfies the equation of limiting equilibrium. The degree of approximation depends on the accuracy of linearisation of the equation of limiting equilibrium. For known stress components, the field of rate of displacement is determined by the equation of viscoplastic flow, i.e.,

$$\left.\begin{aligned}
\dot{\gamma}_i(x,y) &= [\tau_i(x,y) - \tau^*(x,y)]/(2\eta); \\
\tau^*(x,y) &= [\sigma(x,y) - u_w(x,y)]\,\mathrm{tg}\,\varphi + c.
\end{aligned}\right\} \qquad \dots \ (8.17)$$

4. Analysis of Displacements Due to Landslips

In the design, construction and operation of structures interacting with sliding slopes, it becomes necessary to predict the landslip displacements. Such a prediction may be based on an evaluation of the stress-strain of the soil mass, taking into account the rheological properties of soil. At present, there are no efficient methods for solving such problems because the application of numerical methods and the use of computers is required. In the simplest case, creep of slope, it is possible to obtain the solution by analytical methods.

The effect of the variation of pore pressure on the rate of viscoplastic flow of an inclined infinite slope (Fig. 8.2) under a variable pressure acting on its boundary, can be evaluated by examining the problem of propagation inside the slope of a pressure wave emanating from the water-bearing layers lying underneath. This problem was solved earlier [17] for different ratios of the maximum pressure variation on the boundary of a layer to the thickness of the layer, taking into account the creep of soil skeleton, compressibility of pore water and variable loading on the boundary of the layer. For example, if the thickness of the layer is large and the coefficient of permeability of soil is small, then the problem is reduced to the solution of the equation of consolidation (7.21) without considering the initial conditions that satisfy the condition $u_w(0,t) = p\cos\omega t$ at the boundary. This yields the following expression:

$$u_w(y,t) = p\exp\left[-\sqrt{\frac{\omega}{2c_v}}\,y\right]\cos\left[-\sqrt{\frac{\omega}{2c_v}}\,y + \omega t\right]. \qquad \dots \ (8.18)$$

It is evident from the above relation that the amplitude of oscillation of pore pressure decays exponentially, while the phase shift increases linearly with depth of layer.

In the case of a layer of relatively small thickness and a large coefficient of permeability of soil, the boundary pressure $u_w(0,t) = p\cos\omega t$ prevails right

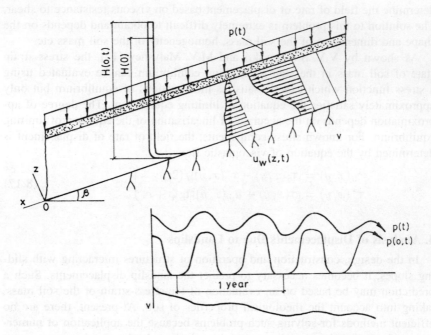

Fig. 8.2. Analytical scheme for prediction of creep deformation of an inclined infinite slope.

across the thickness of the layer. In this case, too, the problem is similarly solved, without considering the boundary conditions. Further, $u_w(h, z) = 0$ and the pressure of stress waves is determined by the following solution:

$$u_w(y, t) = p(y_1 \cos \omega t + y_2 \sin \omega t);$$

$$\left. \begin{array}{l} y_1 = \\ \dfrac{\sin \lambda(h - y)ch\lambda(h - y)\sin \lambda hch\lambda h + \cos \lambda(h - y)sh\lambda(h - y)\cos \lambda hsh\lambda h}{\sin^2 \lambda hch^2\lambda h + \cos^2 \lambda hsh^2\lambda h}; \\[2mm] y_2 = \\ \dfrac{\sin \lambda hch\lambda h \cos \lambda(h - y)sh\lambda(h - y) - \sin \lambda(h - y)ch\lambda(h - y)\cos \lambda hsh\lambda h}{\sin^2 \lambda hch^2\lambda h + \cos^2 \lambda hsh^2\lambda h}. \end{array} \right\}$$

$$\ldots (8.19)$$

If the boundary function is represented as a combination of harmonics of different frequencies, then the solution of the problem may be obtained by superposition of the solutions for individual harmonics. Upon substituting the values of pore pressure obtained from (8.18) and (8.19) in eqn. (8.17), an expression can be derived for rate of displacement of slope which takes into account the variable pore pressure.

The method of prelimit state of stress has been widely used in recent years for prediction of temporal processes occurring in slopes. Unlike the preceding method, in this method it is assumed that limit state can occur only in individual localised regions. This enables the stress-strain state of soil mass to be studied within the framework of linear or non-linear viscoelasticity and the regions of plastic flow to be identified on the basis of stresses. In this case, a complex system of interacting regions is obtained, of which some are in a limit state and others in a prelimit state. The rate of creep is determined not only by the state of stress and its transformation in these regions, but also by the transformations that occur in the limit in space and time. The above problem is so complex that it is extremely difficult to obtain an exact solution even for a homogeneous mass. Therefore, the discussion will be restricted to an approximate method of solution only.

By applying the methods described above for evaluation of the condition of soil mass before attainment of limit state, it will be demonstrated how the variation of strength with time can be taken into account. For example, if the stresses in the soil mass before it attains the limit state are known, then the change in the contour of the regions experiencing plastic flow can be determined on the basis of criterion (8.3), assuming that the variation of the strength parameters with time is known and is given by the following relation:

$$\eta_{str}(t) = \frac{2\tau_{max}\cos\varphi(t)}{c(t) + \mathrm{tg}\,\varphi(t)[\sigma_1 + \sigma_2 - 2\tau_{max}\sin\varphi(t)]}.$$

Thus based on the solution to the elastic problem, it is possible to evaluate as a first approximation the dynamics of the processes of long-term stability occurring in the soil mass, provided the variation of soil strength with time is known from the moment the trench or enbankment was formed.

If, in addition to the gravitational forces, the soil mass is subjected to surface and seepage forces, then the solution to the problem can be obtained in a similar manner by means of the results of the stress-strain state of soil mass described above. While examining the saturated soil mass during the initial stage of formation of the stress-strain state, it is necessary to induct pore pressure in the strength condition (8.3) or carry out the calculations in terms of effective stresses. The pore pressure should be determined by formula (2.63), while the total stresses σ_x and σ_y should be found from the solution of the two-dimensional problem of the theory of elasticity, obtained by the method described above.

Damping creep must be taken into consideration while analysing the initial period of formation of the stress-strain state in the soil mass comprising a young slope. In this case, the displacements and their rates are initially large and can be estimated using the theory of hereditary creep, employing eqn. (8.6), provided it is assumed that the state of stress in an elastic medium experiencing creep is identical to that of the elastic medium. As the initial creep of a slope does not always decay, but may asymptotically approach the stage of creep at constant rate, it becomes necessary to further analyse the rate of displacement of the soil mass. This is possible either by adopting other rheological models or by describing the last stage of steady creep by equations of viscous flow as described below.

While estimating the stress-strain state of soil mass, ideal viscosity is considered in those cases wherein the stability of slope over a given period of time is assured, but slow gravitational displacements disturb the normal service conditions of the structures interacting with the slope. As mentioned earlier, the stress-strain state of soil mass can be evaluated by adopting the incompressible viscous body model, described by eqn. (8.7). By analysing the stress-strain state of soil mass as a plane strain problem, the distribution of strain and displacement rates can be obtained assuming the stress field to be constant. The displacements over a given period of time are determined by formula (8.8).

When the soil mass is inhomogeneous, transient fields of stresses, velocities and strains are produced due to the mutual effect of individual zones of different viscosities. Generally speaking, it is possible to evaluate the above parameters only by numerical methods. However, analytical solutions are also possible in the case of simple contours and areas. For example, multiple-tier landslides can be evaluated in a simple manner by direct integration of equilibrium equations with consideration of the boundary conditions.

9

Numerical Methods for Solving Problems of Consolidation and Creep

1. General Principles

The methods described in the preceding chapters for studying the mechanical properties of soils and determining their rheological parameters, as also the governing equations that describe the relation between stresses and strains can be employed for solution of applied problems by numerical methods. The scope of numerical methods is incomparably wider than that of the analytical methods. This is particularly true in problems dealing with the stress-strain state of non-homogeneous mass, interacting with foundations and structures of finite stiffness. In addition, it is possible by numerical methods to obtain solutions to the problems, taking into account non-linear deformation of soil, its anisotropy etc. Thanks to this feature of numerical methods, it becomes possible to study the stress-strain state of a soil mass for a wide range of variation of stresses, including limit state and prelimit state stresses. This carries great significance in the design of structural foundations since it becomes possible to exploit the untapped reserve of the load-bearing capacity of soil, thereby improving the economic feasibility of the foundation design.

The numerical methods of analysis of the stress-strain state are based on appropriate methods of solution of the differential equations of mathematical physics, developed at the end of the last century and early this century. Soviet scientists made significant contributions to the development of this science. In particular, the Bubnov-Galerkin method of approximate solution of differential equations is being successfully applied in finite element approximation. At present, the numerical methods of solution of applied problems have found maximum application in construction engineering and soil mechanics. These methods have become extremely popular among a wide circle of designers because of their clarity, ease of application for non-homogeneous masses of any geometry and possibility of taking the elasto-viscoplastic behaviour of soil into account.

2. Solution to Problem of Consolidation of Saturated Base

The solution to the problem of consolidation of a saturated base by numerical methods is reduced to the search of stress and strain fields in the soil mass, taking into account the initial and boundary conditions as well as the deformability and permeability of soil. The solution is possible by both the finite difference and the finite element methods.

The crux of the finite difference method lies in the unknown pore pressure function or effective stress being determined only at the nodal points of the grid in steps of Δx_2 inside the region under consideration at instants of time $t_0, t_1, \ldots, t_n$, such that $\Delta t = t_1 - t_0 = t_2 - t_1 = \cdots = \text{const.}$

The solution of differential equations in partial derivatives is reduced to the solution of a system of algebraic equations. Obviously, the accuracy of the numerical solution will increase with decrease of grid step Δx_i and time step Δt, although this will increase the data to be processed and the CPU time required for arriving at the solution.

The numerical methods of solution of the problems of consolidation can be developed on the basis of both explicit and implicit schemes.

The solution to the problem of one-dimensional consolidation by the finite element method, taking into account the compressibility of soil skeleton and pore water and the non-linear behaviour of permeability as well as the geometrical non-linearities described in [17], was developed jointly by the author with I.I. Demin and A.A. Rakhmanov.

The compressibility of soil skeleton is described by the expression:

$$l = l_0 - b\{1 - \exp[-a(\sigma - u_w)]\}, \qquad \ldots (9.1)$$

where a and b are empirical coefficients.

The coefficient of compressibility of pore water can be determined by the formula:

$$m_w = \frac{1 - S_r(1 - \mu)}{u_w + p_a}, \qquad \ldots (9.2)$$

where μ is Henry's coefficient of solubility and p_a is atmospheric pressure.

The coefficient of permeability is likewise a non-linear function of effective stresses:

$$k = k_0 \psi^n, \qquad \ldots (9.3)$$

where

$$\psi = \exp[-a(\sigma - u_w)]. \qquad \ldots (9.4)$$

Taking into consideration (9.4), eqn. (9.1) can be rewritten in the following form:

$$l = l_0 - b(1 - \psi). \qquad \ldots (9.5)$$

Consequently, the equation of one-dimensional consolidation may be written as:

$$\frac{m_w B + (m_v + m_w)\psi}{m_v \psi (A + \psi)} \frac{\partial \psi}{\partial t} = c_v(0) \frac{\partial}{\partial z} \psi^{n-1} \frac{\partial \psi}{\partial z}, \qquad \ldots (9.6)$$

where $A = (1 - l_0 - b)/b$; $B = (l_0 - b)/b$; $c_v^{(0)} = k_0 f_0/(\gamma_w m_v)$.

The following integral substitutions are carried out for solution of (9.6):

$$H(\psi) = \int_{\psi_{min}}^{\psi} \frac{m_w B + (m_v + m_w)\psi}{c_v^{(0)} m_v \psi (A + \psi)} d\psi; \qquad \ldots (9.7)$$

$$\Phi(z) = \int_{\psi_{min}}^{\psi} \psi^{n-1} d\psi = \frac{1}{n}(\psi^n - \psi_{min}^n), \qquad \ldots (9.8)$$

where ψ_{min} is the minimum value of function ψ in the given range of variation of u_w.

Taking into account formulae (9.7) and (9.8), the equation of consolidation (9.6) is written as:

$$\frac{\partial H}{\partial t} = \frac{\partial^2 \Phi}{\partial z^2}. \qquad \ldots (9.9)$$

This equation is valid for the whole region under investigation and the non-linearity is fully incorporated in the expressions for $H(\psi)$ and $\Phi(\psi)$. Consequently, the functions under the symbol of integration may be arbitrary. In view of the foregoing, the initial and boundary conditions become modified as follows:

$t = 0$;

$$\left.\begin{array}{l} H(z,0) = \dfrac{1}{c_v^{(0)}} \ln \left[\left(\dfrac{A+\psi}{A+\psi_{min}}\right)^{1+m_w/m_v} \left(\dfrac{A+\psi_{min}}{A+\psi_1} \dfrac{\psi_1}{\psi_{min}}\right)^{m_w B/(m_v A)} \right]; \\[3mm] \Phi(z,0) = \dfrac{1}{n}(\psi_1^n - \psi_{min}^n); \quad z = 0; \quad H(z,t) = 0; \\[2mm] z = h; \quad \Phi(z,t) = 0. \end{array}\right\}$$

$$\ldots (9.10)$$

Equations of the type (9.9) with the initial and boundary conditions given in the form of (9.10) can be solved by the finite element method. The results for such a solution, taking into account variation in the height of the consolidating layer with time are shown in Fig. 7.3.

The finite element is particularly effective when used in the solution of a two-dimensional problem of consolidation, taking into account non-linear compressibility and permeability of soil as described by the following relations:

$$e = e_0 - b(1 - \psi); \qquad \ldots (9.11)$$

$$k_x = k_x^{(0)} \psi^n; \qquad \qquad \qquad \text{... (9.12)}$$

$$k_z = k_z^{(0)} \psi^n; \qquad \qquad \qquad \text{... (9.13)}$$

$$\psi = \exp\left[-\frac{a}{3}(1+\mu)(\sigma_x^{tot} + \sigma_z^{tot} - u_w)\right],$$

where a, b and n are coefficients determined experimentally; e_0, $k_x^{(0)}$, $k_z^{(0)}$ are initial values of the coefficients of porosity and permeability; σ_x^{tot}, σ_z^{tot} are total values of the normal stress components and u_w is pore pressure.

Considering that $\xi_a = k_x/k_z$, it is found that:

$$k_x = \xi_a k_z \psi^n, \qquad \qquad \text{... (9.14)}$$

where ξ_a is the coefficient of anisotropic permeability determined experimentally.

The equation of consolidation can now be written as follows:

$$\frac{1}{c_v^{(0)}}\left[\frac{1}{A+\psi} + \frac{m_w(B+\psi)}{a(A+\psi)\psi}\right]\frac{\partial\psi}{\partial t} = \xi_a\frac{\partial}{\partial x}\left(\psi^{n-1}\frac{\partial\psi}{\partial x}\right) - \frac{\partial}{\partial z}\left(\psi^{n-1}\frac{\partial\psi}{\partial z}\right),$$

$$\text{... (9.15)}$$

where $c_v^{(0)} = k_z^{(0)}/(\gamma_w m_v)$.

Determination of the boundary conditions is an important step in the solution to the above equation. For fully saturated soil, it is usually assumed that there is no volumetric strain at the beginning and hence Poisson's ratio for the soil skeleton is taken as equal to 0.5. In the case under consideration, when $0.8 \leq S_r \leq 1.0$, the above approach is not applicable because volumetric strain begins to develop as soon as loading is started. It is advisable to apply the method of reduced modulus, according to which in the initial period the behaviour of soil is, on the whole, described by the reduced Poisson's ratio μ_{red} and reduced modulus of deformation E_{red}:

$$\mu_{red} = \frac{K_{red} - 2G}{2(K_{red} + G)};$$

$$E_{red} = 2G(1 + \mu_{red}),$$

where $K_{red} = K + K_w/n$; K is the bulk modulus of soil skeleton under compression; K_w is the modulus of compressibility of pore water; n is porosity and $G = G_{red}$ is the modulus of shear.

If the soil medium is characterised by volumetric incompressibility, i.e., $K_{red} \to \infty$, then $\mu_{red} = 0.5$.

While applying the method of reduced modulus, it is possible to make use of the known solutions of the corresponding boundary value problem of the mechanics of homogeneous medium and determine, first, the distribution of total stresses in the multiphase medium and then the distribution of the pore pressure at the beginning $(t = 0)$:

$$u_w(x, z, 0) = \sigma^{tot}(x, z, 0)A_0, \qquad \dots (9.16)$$

where $\sigma^{tot}(x, z, 0)$ is the initial value of the mean total stress for the reduced medium:

$$\sigma^{tot}(x, z, 0) = \frac{1}{3}(1 + \mu_{red}) \left[\sigma_x^{tot}(x, z, 0) + \sigma_t^{tot}(x, z, 0)\right]. \qquad \dots (9.17)$$

The coefficient of initial pore pressure is:

$$A_0 = \frac{K_w}{K_w + K/n}.$$

In the case of full water saturation $(S_r = 1.0)$ $K_w \to \infty$ hence, $A_0 \to 1$.

In the problem under consideration, it was assumed that the sum of total stresses remains constant in the course of consolidation. The validity of this assumption was demonstrated in [21] by comparing the ratio $(1 + \mu_{red})/(1 + \mu)$, which does not vary more than 20% from the beginning of consolidation to its completion, while μ_{red} and μ change from 0.5 to 0.25 during the same period. For accurate calculations, one may apply the method of successive approximations, deriving the first approximation from the condition which states that the sum of principal total stresses is constant. Thus, the solution to the problem is reduced to the solution of eqn. (9.15) with the following boundary and initial conditions:

$$\psi(x, z, 0) = \exp\left\{-\frac{a}{3}(1 + \mu_{red})[\sigma_x^{tot}(x, z, 0) + \sigma_z^{tot}(x, z, 0)]\right\}(1 - A_0) = \psi_1;$$

$$\psi(x, 0, t) = \psi_2; \quad \psi(0, z, t) = \psi_3.$$

The following integral substitutions are now carried out:

$$H(\psi) = \int_{\psi_{min}}^{\psi} \left\{\frac{1}{c_v^{(0)}} \left[\frac{1}{A + \psi} + \frac{K_w(B + \psi)}{a(A + \psi)\psi}\right]\right\} d\psi;$$

$$\Phi(\psi) = \int_{\psi_{min}}^{\psi} \psi^{n-1} d\psi.$$

With the help of the above substitutions, eqn. (9.15) is reduced to the quasi-linear form given below with the appropriately corrected initial and boundary conditions:

$$\xi_a \frac{\partial^2 \Phi}{\partial x^2} + \frac{\partial^2 \Phi}{\partial z^2} - \frac{\partial H}{\partial t} = 0. \qquad \dots (9.18)$$

In determining the initial and stabilised states of stresses, the relations between stresses and strains are assumed to be non-linear.

—stress-strain relation describing distortion

$$\tau_i = \frac{G_0 \tau_s}{\tau_s + G_0 \gamma_i} \gamma_i, \qquad \qquad \dots (9.19)$$

where G_0 is the initial modulus of shear determined experimentally and τ_s is the limiting value of shearing stress when $\gamma_i \to \infty$;

$$\gamma_i = \sqrt{\frac{2}{3}} \sqrt{(\varepsilon_x - \varepsilon_z)^2 + \frac{3}{2}\gamma^2 xz};$$

—stress-strain relation describing volumetric deformation

$$\sigma'_m = \frac{K_0 \varepsilon_s}{\varepsilon_s - \varepsilon_m} \varepsilon_m, \qquad \qquad \dots (9.20)$$

where K_0 is the initial modulus of volumetric strain (at $\varepsilon_m = 0$) determined experimentally, ε_s is the limiting value of volumetric strain that occurs at $\sigma_m \to \infty$ when the material attains limiting value of density and $\sigma'm$ is the mean effective stress.

Extremum principles have been formulated for the above equations. On the basis of these principles all the relations of the finite element method have been obtained and a computer software package, GEOFLIT has been developed. This package enables:

— Determination of the initial and stabilised stress-strain states of foundations, based on elastic and non-linear elastic models.
— Determination of the initial pore pressure distribution in the foundation and variation of the pore pressure in the course of consolidation.
— Determination of the settlement of structure with time.

Illustrative results of the application of the proposed method after analysis are shown in Figs. 9.1 to 9.5.

3. Solution to the Problem of Creep of Partially Saturated Base[1]

This section discusses the solution to the creep problem by the finite element method in accordance with the earlier works in which this method was widely applied. The equation of state with consideration of time factor is invariably written in the form of a relation between strain rate and stress and strain.

$$\{\dot{\varepsilon}^{vp}\} = \frac{d}{dt}\{\varepsilon^{vp}\} = \beta\left(\{\sigma\}, \{\varepsilon^{vp}\}\right), \qquad \qquad \dots (9.21)$$

where $\{\sigma\}$ is stress vector; $\{\dot{\varepsilon}^{vp}\}$ is the vector of the rate of viscoplastic strain and $\{\varepsilon^{vp}\}$ is the vector of viscoplastic strain.

---

[1] This section was written by A.G. Shchebolev, V.N. Vorobev and N.E. Shakhurina.

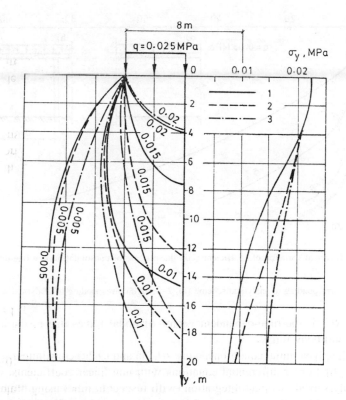

Fig. 9.1. Variation of vertical stresses σ_y, depending on Poisson's ratio.
1—$\nu = 0.35$; 2—$\nu = 0.4$; 3—$\nu = 0.45$.

In the discussion to follow, the viscoplastic strain of the medium will be treated as the creep strain and no distinction will be made between the two. The total strain $\{\varepsilon\}$ in a viscoplastic body is equal to the sum of elastic strain $\{\varepsilon^e\}$ and viscoplastic strain $\{\varepsilon^{vp}\}$ components:

$$\{\varepsilon\} = \{\varepsilon^e\} + \{\varepsilon^{vp}\}. \qquad \cdots \ (9.22)$$

The stresses are related to elastic strain by Hooke's law as follows:

$$\{\sigma\} = [D]\{\varepsilon^e\} = [D](\{\varepsilon\} - \{\varepsilon^{vp}\}), \qquad \cdots \ (9.23)$$

where $[D]$ is elasticity matrix.

The equation of equilibrium can be written as follows in the finite element method:

$$\int_v [B^T]\{\sigma_n\}dv - \{R_n\} = 0; \qquad \cdots \ (9.24)$$

174

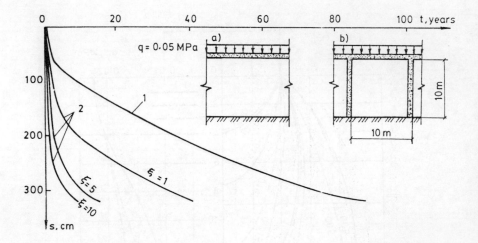

Fig. 9.2. Effect of variation of coefficient ξ on the process of consolidation in a one-dimensional problem.

1—curve corresponding to analytical scheme (a); 2—curve corresponding to analytical scheme (b).

where $\{R_n\}$ is the time-dependent vector of nodal forces due to the action of surface and body forces.

For the given initial conditions, eqns. (9.21) and (9.24) constitute a system of ordinary first-order differential equations with non-linear coefficients, which is invariably solved by direct integration (with respect to time) using the time-step procedure. The solution to the problem is obtained for individual time intervals Δt. Approximation of the time-dependent parameters over Δt is generally carried out by one of the finite difference methods. The choice of the method largely decides the accuracy, stability and economic feasibility of the solution. For increment $\Delta t_n = t_{n+1} - t_n$, the equation of equilibrium is satisfied in incremental form:

$$\int_v [B^T]\{\Delta\sigma_n\}dv - \{\Delta R_n\} = 0, \qquad \dots (9.25)$$

where $\{\Delta R_n\}$ is the increment of nodal force vector due to change in load during time interval Δt_n.

Stress increment $\{\Delta\sigma_n\}$, which depends on the strain increment, is determined as described below. Using the eqn. (9.21). the increment of viscoplastic strain over step Δt_n can be determined by the implicit scheme through the rate of viscoplastic strain at instants of time t_n and t_{n+1} as follows:

$$\{\Delta\varepsilon_n^{vp}\} = \Delta t_n \left[(1 - \Theta)\{\dot\varepsilon_n^{vp}\} + \Theta\{\dot\varepsilon_{n+1}^{vp}\}\right]; \quad (0 \le \Theta \le 1). \qquad \dots (9.26)$$

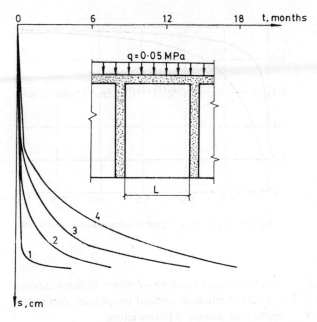

Fig. 9.3. Development of settlement with time as a function of the distance between drains at $\xi = 5$.

1—$L = 1$ m; 2—$L = 2$ m; 3—$L = 3$ m; 4—$L = 4$ m.

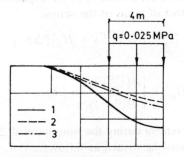

Fig. 9.4. Distribution of the settlement of foundation under the action of external load.

1—stabilised settlement; 2—initial settlement; 3—settlement due to distortion.

Relation (9.26) is based on the assumption of linear variation of the rate of viscoplastic strain in the time interval between t_n and t_{n+1}. Depending on the selected value of Θ, integration can be carried out by one of the following

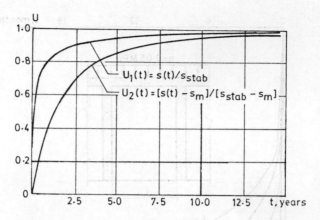

Fig. 9.5. Variation of degree of consolidation u with time.

methods:

$\Theta = 0 -$ Euler's method (method of direct differentiation);
$\Theta = 1/2 -$ Crank-Nicholson method (trapezium method);
$\Theta = 1 -$ method of inverse differentiation

Euler's method is a method of explicit integration because the increment of viscoplastic strain is determined from the stress-strain state at instant t_n. In the Crank-Nicholson and inverse differentiation methods, the plastic strain $\dot{\varepsilon}^{vp}_{n+1}$ in relation (9.26) is integrated for instant t_{n+1} by expanding into Taylor's series, retaining a limited number of terms of the series:

$$\{\dot{\varepsilon}^{vp}_{n+1}\} = \{\dot{\varepsilon}^{vp}\} + [H_n]\{\Delta\sigma_n\}, \qquad \ldots (9.27)$$

where

$$[H_n] = \left[\frac{\partial\{\dot{\varepsilon}^{vp}_n\}}{\partial\{\sigma\}}\right]_n = [H_n(\{\sigma_n\})]; \qquad \ldots (9.28)$$

$\{\Delta\sigma_n\}$ is the stress increment during the time interval Δt.
Relation (9.26) can now be written as follows:

$$\{\Delta\varepsilon^{vp}_n\} = \{\dot{\varepsilon}^{vp}_n\}\Delta t + [C_n]\{\Delta\sigma_n\}; \qquad \ldots (9.29)$$

$$[C_n] = \Theta\Delta t_n[H_n]. \qquad \ldots (9.30)$$

Stress increment $\{\Delta\sigma_n\}$ is equal to the product of elastic strain increment $\Delta\varepsilon^e_n$ and the elasticity matrix:

$$\{\Delta\sigma\} = [D]\{\Delta\varepsilon^e_n\} = [D](\{\Delta\varepsilon_h\} - \{\Delta\varepsilon^{vp}_n\}). \qquad \ldots (9.31)$$

The increment of total strain can be expressed through the increment of nodal displacements by the relation:

$$\{\Delta\varepsilon_n\} = [B]\{\Delta d_n\}. \qquad \qquad \ldots (9.32)$$

Substituting $\{\Delta\varepsilon_n^{vp}\}$ from (9.29) in eqn. (9.31), the following expression is obtained for stress increment.

$$\{\Delta\sigma_n\} = [\hat{D}_n]([B]\{\Delta d_n\} - \{\dot{\varepsilon}_n^{vp}\}\Delta t), \qquad \ldots (9.33)$$

where, $[\hat{D}_n] = [D]^{-1} + [C_n]^{-1}$.

For the explicit integration scheme ($\Theta = 0$), eqn. (9.33) can be simplified and written as follows:

$$\{\Delta\sigma_n\} = [D]([B]\{\Delta d_n\} - \{\dot{\varepsilon}_n^{vp}\}\Delta t_n). \qquad \ldots (9.34)$$

Thus a closed form expression is obtained for stress increment as a function of the total stress increment during interval Δt_n and the rate of viscoplastic strain at instant t_n. Expression (9.33) for stress increment is used directly in the equilibrium equation. Now, applying relations (9.29) and (9.33), the following equation is obtained:

$$[K_{T,n}]\{\Delta d_n\} - \int_v [B^T][\hat{D}_n]\{\dot{\varepsilon}_n^{vp}\}\Delta t_n dV - \{\Delta R_n\} = 0, \qquad \ldots (9.35)$$

where $[K_{T,n}] = \int_v [B^T][\hat{D}_n][B]dV$ is the tangential stiffness matrix.

Equation (9.35) can be rewritten in the conventional form:

$$\{\Delta d_n\} = [K_{T,n}]^{-1}\{\Delta V_n\}; \qquad \ldots (9.36)$$

where $\{\Delta V_n\}$ is the increment vector of fictitious nodal forces due to initial viscoplastic strain:

$$\{\Delta V_n\} = \int_v [B^T][\hat{D}_n]\{\dot{\varepsilon}_n^{vp}\}\Delta t_n dV + \{\Delta R_n\}. \qquad \ldots (9.37)$$

Upon solving eqn. (9.36) we can determine the increment of nodal displacements, which, upon substitution in formula (9.33) yields the stress increment $\{\Delta\sigma_n\}$. Hence,

$$\{\sigma_{n+1}\} = \{\sigma_n\} + \{\Delta\sigma_n\}; \qquad \ldots (9.38)$$
$$\{d_{n+1}\} = \{d_n\} + \{\Delta d_n\}. \qquad \ldots (9.39)$$

With the help of relations (9.31) and (9.32), the increment of viscoplastic strain $\{\Delta\varepsilon_n^{vp}\}$ can be determined as follows:

$$\{\Delta\varepsilon_n^{vp}\} = [B]\{\Delta d_n\} - [D]^{-1}\{\Delta\sigma_n\}, \qquad \ldots (9.40)$$
$$\{\varepsilon_{n+1}^{vp}\} = \{\varepsilon_n^{vp}\} + \{\Delta\varepsilon_n^{vp}\}. \qquad \ldots (9.41)$$

Stress increment $\{\Delta\sigma_n\}$ is determined on the basis of the solution of equilibrium equation (9.25) in linearised form. Therefore, stress $\{\sigma_{n+1}\}$ determined on the basis of stress increment from formula (9.38) may not satisfy the equilibrium equation (9.24). A number of computational procedures are available for correcting the value of σ_{n+1}. One of these procedures has been realised in the form of a computer program developed by A.G. Shchebolev to solve the problem of creep of frozen soil. The underlying principle of this procedure is that each time step is checked to ascertain whether equilibrium equation (9.24) is satisfied. If it is not satisfied, then the discrepancy vector of the nodal forces is determined for instant t_{n+1} and added to the vector of fictitious nodal forces determined by formula (9.37) in the next time step. The discrepancy vector is given by the following relation:

$$\{R^*_{n+1}\} = \int_v [B]^T \{\sigma_{n+1}\}dV - \{R_{n+1}\} \neq 0. \qquad \ldots (9.42)$$

This computational procedure allows refinement of the solution, on the one hand, and avoids the iterative procedure during the time interval Δt_n, on the other.

An important step in the solution of particular problems concerns the selection of time interval Δt since it largely determines the accuracy of the solution and its stability. The integration scheme is considered stable if the error of integration does not increase progressively during the calculations and the solution is free of any undamped oscillations. At $\Theta < 1/2$ the integration scheme is conditionally stable, i.e., it is stable if step Δt is less than a certain critical value. In Euler's explicit integration scheme ($\Theta = 0$), time interval Δt is restricted for the particular type of governing equations. In actual practice, step Δt is selected on the basis of empirical relations that are obtained as a result of numerical experiments. When $\Theta \geq 1/2$, the integration scheme is unconditionally stable, i.e., there are no restrictions on the time interval Δt. However, in both the conditionally and unconditionally stable schemes, the accuracy of the solution obtained depends to a large extent on step Δt. Therefore, this factor imposes certain restrictions on the magnitude of Δt. In solving creep problems there is no relation between temporal and spatial discretisations, i.e., the selection of step Δt, does not depend on the size of the elements used for approximation of the region under consideration.

In solving creep problems it is possible to utilise constant step as well as variable step Δt. In the latter case, it is possible to strike a better compromise between the requirements of economy, accuracy and stability of the solution. Step Δt is determined during the computational procedure by established empirical relations, depending on the anticipated variation of viscoplastic strain. Obviously, interval Δt may increase considerably as the stabilised state is approached.

Step Δt is determined from the condition that the maximum increment of the viscoplastic strain should not be in excess of a certain fraction of the total strain:

$$\Delta \varepsilon_{n(i)}^{vp} = \varepsilon_{n(i)}^{vp} \Delta t_n \le \tau \varepsilon_n(i), \qquad \ldots (9.43)$$

or

$$\Delta t_n \le \tau \left[\frac{\varepsilon_{n(i)}}{\dot{\varepsilon}_{n(i)}^{vp}} \right]_{min}^{1/2}, \qquad \ldots (9.44)$$

where $\varepsilon_{n(i)}$ is strain and $\dot{\varepsilon}_{n(i)}^{vp}$ is the rate of viscoplastic strain.

For isoparametric elements, condition (9.43) is checked at all points of integration by the Gauss criterion and the minimum value of Δt is used in the solution. For the explicit schemes, coefficient τ is selected in the range $0.01 \le \tau \le 0.15$, while for the implicit schemes it may be much higher, subject to the condition $\tau \le 10$.

The other criterion which is employed for determining Δt_n imposes restrictions on the difference in magnitude of the step obtained from successive approximations:

$$\Delta t_n \le K \Delta t_{n-1}, \qquad \ldots (9.45)$$

where K is a coefficient that is always taken as equal to 1.5.

Some researchers suggest that Δt should be determined from the condition that the increment of viscoplastic strain should not exceed one-half of the elastic strain:

$$\{\Delta \varepsilon_n^{vp}\} = \{\dot{\varepsilon}_n^{vp}\} \Delta t \le \frac{1}{2} \{\varepsilon_n^e\}. \qquad \ldots (9.46)$$

The process of solving the problem continues until a stabilised state is attained or the time limit is reached. Whether the stabilised state has been attained by the numerical solution is judged from the increment of viscoplastic strain after each time step. The rate of viscoplastic strain can serve as the convergence criterion. In this case, the stabilised state is said to have been attained at the end of a step if

$$\frac{\Delta t_{n+1} \Sigma \dot{\varepsilon}_{(l)n+1}^{vp}}{\Delta t_1 \Sigma \dot{\varepsilon}_{(i)1}^{vp}} 100 \le \epsilon, \qquad \ldots (9.47)$$

where ϵ is the given accuracy of approximation to the stabilised state, which is taken as $\epsilon = 1$ (i.e., 1%).

The solution of expression (9.47) yields an integrated index that serves as a global convergence criterion. In changing over to local convergence, condition (9.46) must be satisfied at each point of the Gaussian integration in which viscoplastic deformation is observed.

Let us examine the stress-strain state of the foundation of a heavy structure for the case when the soil of the base experiences viscoplastic flow. The analysis will be based on the solution of the problem by the finite element method as derived by V.N. Vorobev. It is assumed that for elasto-viscoplastic soil the tensor of strain increment can be represented in the following form:

$$d\varepsilon_{ij} = d\varepsilon_{ij}^e + d\varepsilon_{ij}^{vp}, \qquad \dots (9.48)$$

where $d\varepsilon_{ij}^e$ and $d\varepsilon_{ij}^{vp}$ represent the components of elastic and viscoplastic strain increment respectively.

The components of elastic strain increments are determined on the basis of the genralised Hooke law, while the components of viscoplastic strain increments are found on the basis of the V. Koiter's generalised associated law:

$$d\varepsilon_{ij}^{vp} = \sum_r \Delta th_r \frac{\partial f_r^{(0)}}{\partial \sigma_{ij}}, \qquad \dots (9.49)$$

where $\Delta th_r = d\lambda_r$ is an indefinite scalar multiplier, r represents regular sections of the loaded surface and σ_{ij} represents components of the working stress tensor.

The tensor of active stresses $\hat{\sigma}_{ij}$ is introduced in place of the tensor of working stresses σ_{ij} in order to take into account the effect of time lapsed since the application of load:

$$\hat{\sigma}_{ij} = \sigma_{ij} - r_{ij}, \qquad \dots (9.50)$$

where r_{ij} represents the components of stress correction tensor which depends, in general, not only on the rate of viscoplastic strain, but also its magnitude.

Within the framework of the theory of viscoplastic hardening, it is assumed that there exists an instantaneous loading surface, which is defined by an equation similar to that of the stabilised load function, but whose arguments are invariants of the active stress tensor:

$$f_r^{(0)} = \hat{\sigma}_i + h_r\hat{\sigma} - \tau_r; \quad r = 1, 2, \dots \qquad \dots (9.51)$$

The model based on eqns. (9.48) to (9.51) [5] has been employed for prediction of the settlement of a heavy foundation under plane strain conditions by the method of elements. The parameters of the model were determined from the results of three-dimensional testing of remoulded clayey soil specimens. The specimens were taken from the foundations of real structures. The parameters determined under axisymmetrical state of stress [5, 6] were recalculated by a practical approximation technique to make them valid for plane strain conditions. The foundation was considered to be homogeneous and it was assumed that the process of its deformation under load was governed by the elastoplastic model described above.

Following the adopted analytical scheme (Fig. 9.6), the settlement of foundation under the load of heavy structure was calculated by taking into account the previous loading history. By the time the deformation process had stabilised, the maximum settlement at the centre of the structure comprised 58 cm, whereas the mean settlement of the surface under load was 44 cm. The inclination of the foundation in the stabilised state was 0.0022, which was within the permissible limits for such structures (Fig. 9.7).

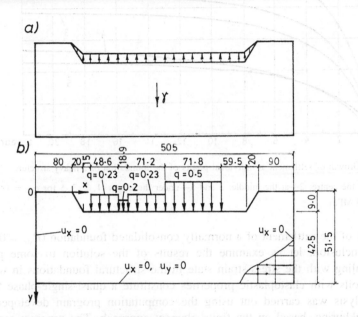

Fig. 9.6. Analytical schemes for determining the settlements of a foundation by the finite element method (dimensions in m; load in MPa).

a—after excavation dug; b—after the structure has been raised.

In evaluating the settlement of foundations, it should be kept in mind that the analysis is carried out for plane strain conditions, the strength characteristics of soil are determined on remoulded specimens and the load is applied in the last stage instantaneously and according to the flexible scheme, although actually the rate of application of load on the foundation would be less and would, therefore, undoubtedly produce lesser settlement during the construction period. In the case of instantaneous load application, about 75% of the final stabilised settlement occurs during the first 4 years. It may be noted that all the factors listed above result in larger settlement, i.e., the above analysis gives the upper bound estimate of the settlement of foundation.

In addition to the foregoing, mention must also be made of the significant effect of the natural condition of the foundation. For example, the final settlement (curve 4, Fig. 9.7.) of an over-consolidated foundation ($P_{str} = 0.15$ MPa) is less

182

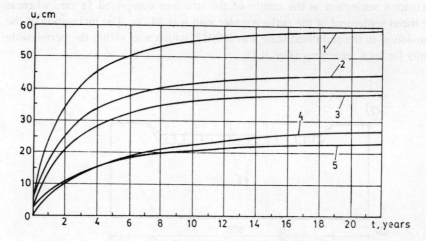

Fig. 9.7. Growth of settlement at individual points of the foundation of a heavy structure.

1 to 4—at the centre; 2—at the middle; 3 and 5—under the edges (1, 2, 3, 5 for $p_{str} = 0$; 4 for $p_{str} = 0.15$ MPa).

than half of the settlement of a normally consolidated foundation ($P_{str} = 0$).

In conclusion, let us examine the results of the solution to some problems dealing with the stress-strain state in the structural foundations in which clayey soils with elastoplastic properties constitute a quasi-single-phase base. The analysis was carried out using the computation program developed by N.E. Shakhurina, based on the finite element approach. The program realises the incremental model of deformability of clayey soil described in Chapter 2 and is within the framework of the plane strain problem. The parameters of the model [see formulae (2.47)] determined from the results of triaxial tests are given in Table 9.1 The discretisation of the regions by finite elements and the geometrical dimensions are shown in Fig. 9.8.

Table 9.1.

Parameter	Numerical value	Parameter	Numerical value
K, MPa	319	n_1, MPa^{-1}	0.35
G, MPa	111.11	n_2, MPa^{-1}	2.162
m_1	0.12	n_3, MPa^{-1}	1.136
m_2	0.812	c, MPa	0.42
m_3	0.65	φ, deg	47

Note: Indices 1, 2 and 3 correspond to parameters relating to loading, unloading and repetitive loading.

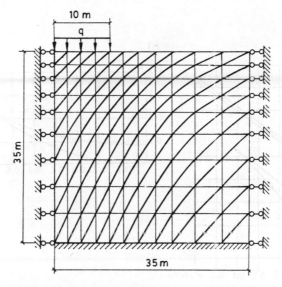

Fig. 9.8. Discretisation of regions subjected to analysis by finite elements.

We examine here a homogeneous layer of clayey soil on a bed in the form of an incompressible base at a depth of 35 m. No slip condition was assumed to occur at the lower boundary $(u = v = 0)$. It was further assumed that there was no sideways displacement of the axis of symmetry and the side boundary of the region under consideration. The algorithm and computer program were checked by means of a mathematical experiment, simulating the testing of the soil specimen under consolidation. The trajectory of compressive loading of the soil specimen obtained by the numerical experiment is shown in Fig. 9.9(a).

It can be seen from the experimental results that when a layer of soil was subjected to constrained compression (Fig. 9.9(b)) followed by total removal of load, the soil mass retained horizontal stresses which accumulated each time the loading-unloading cycle was repeated. These stresses result in an increase in the coefficient of lateral pressure that, in turn, leads to a reduction of the settlement of the structure (Fig. 9.10).

The stress-strain state of the foundation under the action of external load was determined by considering the following two possible initial states of stress; in the first version it was assumed that the initial stress-strain state formed under the self-weight of the soil under conditions of constrained compression with a uniformly distributed load acting on the surface of the foundation; in the second version it was assumed that the initial stress-strain state formed under the action of self-weight only. In the third version, the initial stress-strain state was not taken into consideration in the analysis.

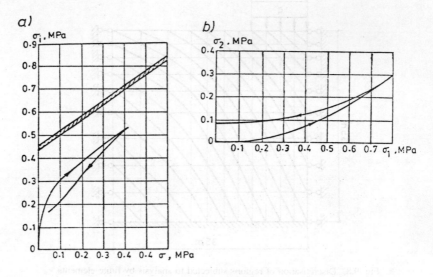

Fig. 9.9. Constrained compression of soil specimen.

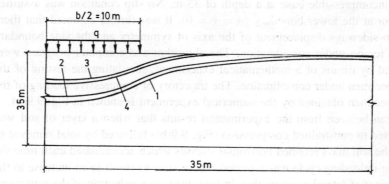

Fig. 9.10. Settlement of the surface of soil layers subjected to flexible load $q = 0.6$ MPa.

1—taking into account the initial state of stress due to self-weight of soil; 2—without taking into account the initial state of stress; 3—taking into account the initial state of stress due to self-weight of soil and additional load $q = 0.4$ MPa; 4—taking into account the initial state of stress due to additional load $q = 0.4$ MPa.

Figure 9.8 depicts the discretisation of the region subjected to analysis under the load of a strip foundation of finite stiffness. Conditions of no slip were assumed to prevail at the interface. A uniformly distributed force p was applied on the plate. The material of the plate was assumed to be elastic with its modulus of elasticity corressponding to that of concrete.

From the analytical results presented in Figs. 9.11 and 9.12, it is evident that the variation of the deformation of the soil base across the depth follows a triangular distribution. The settlement of the base under a plate of finite stiffness is slightly less than that under a flexible plate.

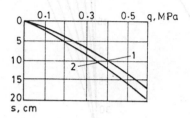

Fig. 9.11. Variation of the settlement of a structure with load.

1—plate of finite stiffness; 2—flexible load.

4. Effect of Type of Initial Stress-strain State of Soil Base on its Settlement under Weight of Structure

An analysis of the methods of prediction of the settlement of foundations indicated that one of the main factors affecting settlement is the initial stress-strain state of the soil base of the structure. At present, the effect of this factor is taken into account by means of one of the components of the initial stress-strain state, i.e., $\sigma_z = \gamma_z$. It has been established by experimental and theoretical studies conducted in recent years that if all components of the initial stress-strain state are taken into consideration, this would greatly increase the accuracy with which the settlement of foundations of structures is determined. This applies first of all to consideration of the initial (before the structure is raised) coefficient of lateral pressure $\xi_0 = \sigma_x/\sigma_z$. It is known that this coefficient may vary between 0.3 and 1.5 and more, depending on the conditions under which the soil mass is formed and the method by which the base is prepared (excavation of soil from pit, dewatering, ramming, installation of soil piles for creating excess lateral pressure etc.).

In view of the foregoing, by the time the foundation is raised, the stress ellipsoid may have differing orientations with respect to the vertical axis of z, i.e., the major axis may be vertical or horizontal (Fig. 9.13). The Nadai-Lode parameter may be selected as the criterion describing the initial stress-strain state of soil

$$\mu_\sigma = (2\sigma_2 - \sigma_1 - \sigma_3)/(\sigma_1 - \sigma_3),$$

where $\sigma_1 > \sigma_2 > \sigma_3$.

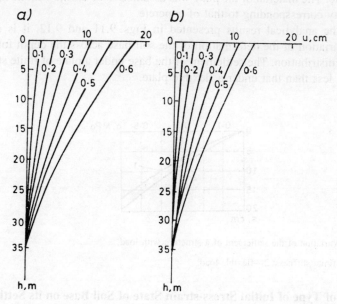

Fig. 9.12. Diagram depicting vertical displacement of plate of finite stiffness in the course of loading.

a—under centre of the plate; b—under edge of the plate.

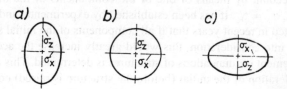

Fig. 9.13. Initial stress-strain state of soil base at different values of the coefficient of lateral pressure.

a—$\xi_0 = 0.5$; b—$\xi_0 = 1$; c—$\xi_0 = 2$.

Here, $\mu_\sigma = -1$ at $\xi_0 < 1$, $\mu_\sigma = 1$ at $\xi_0 > 1$ and $\mu_\sigma = 0$ at $\xi_0 = 1$. This indicates that the additional load on the soil base due to weight of the structure will change the orientation of the major axis of the stress ellipsoid with respect to the z-axis (the foundation axis) and will also result in a change in parameter μ_σ. Consequently, this will lead to different loading trajectories and different deformations of the soil base, depending on the initial stress-strain state.

It is evident from Fig. 9.14 that minimum settlement occurs at $\xi_0 = 0.5$ and

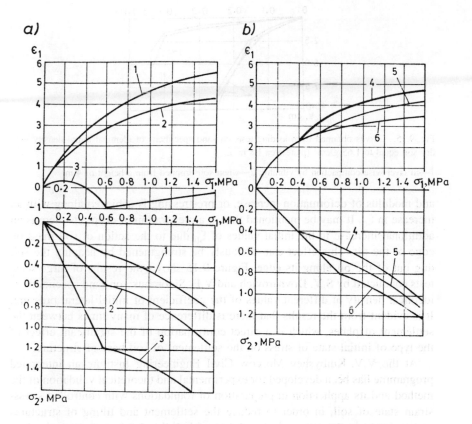

Fig. 9.14. Curves obtained from compression testing of specimens of clayey soil with an initial density of soil particles of 1.96 g/cm^3 (data from N.M. Mkrtchyan).

a—at different values of coefficient of initial lateral pressure ξ_0 and $\sigma_z = 0.6$ MPa (1—at $\xi_0 = 0.5$, 2—at $\xi_0 = 1$, 3—at $\xi_0 = 2$); b—at different values of initial stress σ_z and $\xi_0 = 1$ (4—at $\sigma_z = 0.4$ MPa, 5—at $\sigma_z = 0.6$ MPa, 6—at $\sigma_z = 0.8$ MPa).

maximum at $\xi_0 = 2$. At $\xi_0 = 1$, the settlement vs load curve occupies an intermediate position between the two. However, what is most important is that at $\xi_0 = 0.5$ the deformation of the soil specimen is many times that at $\xi_0 = 2$ (see Fig. 9.14a). It follows, therefore, that the settlement can be predicted with much higher accuracy if the initial stress-strain state of soil is taken into consideration. If necessary, the anticipated settlement can be manipulated by artificially increasing the initial coefficient of lateral pressure. The most effective methods in this regard are consolidation of soil base with heavy or deep compaction using soil piles. This produces two effects: either density of the soil skeleton

188

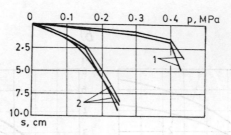

Fig. 9.15. Curves obtained from loading tests on sand specimens of identical initial density with dry soil equal to 1.65 g/cm³ (plate area 1.8 m²).

1—in layer-wise compaction by vibrator; 2—when soil dropped from a height of 4 m.

and modulus of deformation increase, or prestressing is achieved along with an increase in ξ_0. It may be mentioned that in a preparation of deep pits there is an additional increase in the initial values of ξ_0 due to the action of soil mass on edge of the pit. This effect can obviously be strengthened by heaping the soil dug out of the pit along its edge. Figure 9.15 shows the results of large-scale tests conducted by S.V. Dovnarovich and V.F. Sidorchuk[2] on sand specimens of identical density at different values of the coefficient of initial lateral pressure. It is evident from the results that there is difference of many times between the settlement of plates, which is a direct confirmation of the significant effect of the type of initial state of stress on the settlement of soil base of structure.

At the V.V. Kuibyshev Moscow Civil Engineering Institute an integrated programme has been developed for experimental and theoretical validation of the method and its application in preparation of foundations with controlled stress-strain state of soil, in order to reduce the settlement and tilting of structures and to improve the economic feasibility of building heavy structures in deep excavations.

[2] S.V. Dovnarovich, D.E. Pol'shin, D.S. Baranov and V.F. Sidorchuk. Effect of the nature of formation of base on its stressed state. *Osnovaniya, Fundamenty i Mekhanika Gruntov*, No. 6 (1977).

References

1. Vyalov, S.S. 1978. Reologicheskie osnovy mekhvnikigruntov (Rheological Principles of Soil Mechanics). Vysshaya Shkola, Moscow, 447 pp.
2. Gol'din, A.L. and L.N. Rasskazov. 1987. Proektirovanie gruntovykh plotin (Design of Earth Dams). Energoatomizdat, Moscow, 304 pp.
3. Gol'dshtein [Goldstein], M.N. 1973. Mekhanicheskie svoistva gruntov (Mechanical Properties of Soils). Stroiizdat, Moscow, 374 pp.
4. Gorelik, L.V. 1975. Raschety konsolidatsii osnovanii i plotin iz gruntovykh materialov (Analysis of Consolidation of Foundations of Earth Dams). Energiya, Moscow, 154 pp.
5. Zaretskii, Yu.K. and V.N. Lombardo. 1983. Statisktika i dinamika gruntovykh plotin (Statics and Dynamics of Earth Dams). Energiya, Moscow, 256 pp.
6. Zaretskii, Yu.K. 1988. Vyazkoplastichnost' gruntov i raschety sooruzhenii (Viscoplasticity of Soils and Design of Structures). Stroiizdat, Moscow, 350 pp.
7. Zenkevich [Zenkowics], O. and K. Morgan. 1986. Konechnye elementy i approksimvtsiya (Finite Elements and Approximation). Mir, Moscow, 318 pp.
8. Ivanov, L.L. 1985. Grunty i osnovaniya gidrotekhnicheskikh sooruzhenii (Soils and Foundations of Hydrotechnical Structures). Vysshaya Shkola, Moscow, 352 pp.
9. Kul'chitskii, L.I. and O.G. Us'yarov. 1981. Fiziko-khimicheskie osnovy formirovaniya svoistv glinistykh porod (Physical and Chemical Principles in Development of Properties of Clays). Nedra, Moscow, 178 pp.
10. Lavrent'ev, M.A. and B.V. Shabat. 1973. Metody teorii funktsii kompleksnogo peremennogo (Methods of the Theory of Functions of Complex Variables). Nauka, Moscow, 736 pp.
11. Lekhnitskii, S.G. 1977. Teoriya uprugosti anizotropnogo tena (Theory of Elasticity of Anisotropic Solids). Nauka, Moscow, 415 pp.
12. Malyshev, M.V. 1980. Prochnost' gruntov i ustoichvost' osnovanii sooruzhenii (Strength of Soils and Stability of Foundations of Structures). Stroiizdat, Moscow, 136 pp.
13. Maslov, N.N. 1982. Osnovy inzhenernoi geologii i mekhaniki gruntov (Principles of Engineering Geology and Soil Mechanics). Vysshaya Shkola, Moscow, 511 pp.

190

14. Meschyan, S.R. 1978. Nachal'naya i dlitel'naya prochnost' glinistykh gruntov (Initial and Long-term Strength of Clayey Soils). Nedra, Moscow, 207 pp.
15. Nadai, A. 1969. Plastichnost' i razrushenie tverdykh tel (Plasticity and Failure of Solids). Mir, Moscow, 863 pp.
16. Sergeev, E.M. 1982. Inzhenernaya geologiya (Engineering Geology). MGO, Moscow, 240 pp.
17. Ter-Martirosyan, Z.G. 1986. Prognoz mekhanicheskikh protsessov v massivakh mnogofaznykh gruntov (Prediction of Mechanical Processes in Multiphase Soil Masses). Nedra, Moscow, 292 pp.
18. Tikhonov, A.N. and A.A. Samarskii. 1973. Uravneniya matematicheskoi fiziki (Equations of Mathematical Physics). Nauka, Moscow, 724 pp.
19. Fadeev, A.B. 1987. Metod konechnykh elementov v geomekhanike (Finite Element Methods in Geomechanics). Nedra, Moscow, 221 pp.
20. Florin, V.A. 1959, 1961. Osnovy mekhaniki gruntov (Principles of Soil Mechanics). Stroiizdat, Moscow-Leningrad, vol.1 (1959), 357 pp.; vol. 2 (1961), 544 pp.
21. Tsytovich, N.A. and Z.G. Ter-Martirosyan. 1981. Osnovy prikladnoi geomekhaniki v stroitel'stve (Principles of Applied Geomechanics in Construction). Vysshaya Shkola, Moscow, 317 pp.
22. Tsytovich, N.A. 1983. Mekhanika gruntov (Soil Mechanics). Vysshaya Shkola, Moscow, 288 pp.

Printed in India